T0211554

SpringerBriefs in Statistics

More information about this series at http://www.springer.com/series/8921

George Tambouratzis · Marina Vassiliou
Sokratis Sofianopoulos

Machine Translation
with Minimal Reliance
on Parallel Resources

 Springer

George Tambouratzis
Institute for Language and
   Speech Processing
Athens
Greece

Sokratis Sofianopoulos
Institute for Language and
   Speech Processing
Athens
Greece

Marina Vassiliou
Institute for Language and
   Speech Processing
Athens
Greece

ISSN 2191-544X
SpringerBriefs in Statistics
ISBN 978-3-319-63105-9
DOI 10.1007/978-3-319-63107-3

ISSN 2191-5458   (electronic)

ISBN 978-3-319-63107-3   (eBook)

Library of Congress Control Number: 2017947698

Printed on acid-free paper

This Springer imprint is published by Springer Nature
The registered company is Springer International Publishing AG
The registered company address is: Gewerbestrasse 11, 6330 Cham, Switzerland

Acknowledgements

The work presented in this book has originated from the PRESEMT project, which has comprised in total six partners, namely ILSP (Institute for Language and Speech Processing/Athena R.C.), GFAI (Gesellschaft zur Förderung der Angewandten Informationsforshung e.V.), NTNU (Norges Teknisk-Naturvitenskapelige Universitet), ICCS (Institute of Communication and Computer Systems), MU (Masaryk University) and LCL (Lexical Computing Ltd.).

The concept of the PRESEMT methodology was conceived within the Machine Translation Department of ILSP, in a collaborative effort during early 2009. The novelty of the concept is that it attempts to circumvent the requirement for specialised resources and tools so as to support the creation of MT systems for diverse language pairs without constraints. The authors wish to acknowledge within this process the contribution of other members of the department. In addition, we would like to note the contribution of the late Prof. George Carayannis, who was serving as the Director of ILSP and had a key role when starting the work to define the PRESEMT project. His presence is still felt in ILSP.

Though the work described in this book represents the work carried out at ILSP, the authors wish to acknowledge the substantial input of other partners in the creation of the final prototype, in terms of both the different modules and resources. Also, the authors wish to acknowledge the contribution of the late Adam Kilgarriff, Founder of LCL, with whom the Machine Translation Department of ILSP had a close association over more than 10 years. Adam Kilgarriff contributed to the implementation of PRESEMT in terms of assembling language resources.

PRESEMT was selected for funding by the European Commission in September 2009, within FP7/Call4. This funding has been instrumental in performing the related work, and the authors wish to acknowledge the contribution of this financial support (and the support of European Commission project officers as well as external reviewers) in completing the work summarised in this book.

Contents

Chapter 1
Preliminaries

This chapter contains a general introduction to the topic of the present book. It presents the current challenges of Machine Translation (MT), in particular for languages where only a limited amount of specialised resources is readily available. To that end, a comprehensive review of the state-of-the-art in MT is performed. Focus is placed on related work on MT methodologies that are portable to new language pairs, and issues such as stability and extensibility are emphasised. It is widely accepted that language portability necessitates an algorithmic approach to extract information from large corpora in an unsupervised manner. This includes both Statistical MT (SMT) and Example-based MT (EBMT). Here, a review of the strengths and shortcomings of the different approaches is performed, in terms of the a priori externally-provided linguistic knowledge and required specialised resources. This review leads to the concept of the proposed MT methodology.

1.1 Challenges in MT—Relevance to the European Environment

Machine Translation is a term covering the use of computers to automatically (i.e. without the intervention of humans) transform a given text from one language (termed the source language—SL) to another (termed the target language—TL). If a given text is expressed in two languages, the corresponding SL and TL texts can be substantially different both structurally and lexically (Jurafsky and Martin 2009: Chap. 25). Therefore, an effective translation system will need to successfully implement the appropriate translation decisions in both structural (i.e. issues concerning the structure of the translation) and lexical levels (i.e. the choice of words to translate each of the input sentence words). In a human language, a given concept or meaning can be expressed in many ways, and thus different SL texts can have the same desired TL equivalent, complicating the translation task.

© The Author(s) 2017
G. Tambouratzis et al., *Machine Translation with Minimal Reliance on Parallel Resources*, SpringerBriefs in Statistics,
DOI 10.1007/978-3-319-63107-3_1

This level of complexity justifies why, though MT has been researched since the mid-1950s, it remains an active field for research. In addition, though MT systems have now been developed and are commercially available (or also freely available via open-to-use systems) for language pairs involving frequently-used languages, the situation is much less promising when focusing on lesser spoken languages. In the modern multilingual society, the number of combinations of source and target language pairs is very large. For instance, in the European Union, only taking account the official languages of member states (currently 24 official languages), the number of source and target language combinations reaches 552 (= 24 × (24 − 1)). If one also takes into account the semi-official languages and minority languages, the number of combinations rises to much higher levels. Creating functioning MT systems for all of these combinations can easily become an insurmountable problem, yet this remains a requirement in the digital European environment.

The EU has repeatedly stated the aim to support all European languages, with the rationale that each language represents an important part of the cultural heritage of Europe. Each European citizen needs to be able to access information in his/her native language. As a result, the European Commission in 2009 opened a call for project proposals focusing on the development of new MT systems stating that:

> With 23 official languages, the EU is at the forefront of multilingualism and it would be unrealistic to assume that the lingua franca in machine translation is, or will remain, English. A strategic challenge for Europe in today's globalised economy is to overcome language barriers through technological means.

A number of impacts were listed in this call, including the following key ones:

- "Practical and economically viable solutions for fully automatic provision of multilingual online content and services for the vast majority of EU languages";
- "Novel language and translation models that support self-improving, knowledge-driven and interactive paradigms."

As a response to these specified impacts, a research project was proposed to the European Commission under the title "PRESEMT: Pattern Recognition-Based Statistically Enhanced MT". This proposal was subsequently selected for funding under the PRESEMT project (FP7-CALL4, ICT-248307). This activity focused on creating a novel methodology for Machine Translation based predominantly on pattern recognition principles operating on large collections sets of inexpensive data. Thus, the resulting methodology will (i) be readily portable to new language pairs and (ii) require very few specialised resources, potentially sacrificing some accuracy in terms of translation quality to allow for a much simplified process to create a working system for a new language pair, requiring fewer specialised resources. Based on this set of specifications, the MT methodology presented in this volume was designed and developed, with the financial support of the European Commission.

1.2   A Brief Review of MT Development History

Originally, the main impetus to MT research is attributable to the efforts of the US agencies to translate texts from foreign languages of interest (initially Russian) to English. Initial hopes for rapid progress in the field of MT towards automatically produced high-quality translations were dashed by a report on the usability of MT, compiled by the Automatic Language Processing Advisory Committee (ALPAC). This report reached the conclusion that MT was of limited value (for a concise review of the report findings and its effects cf. Hutchins 1996). A major consequence of the ALPAC report was that MT funding became much more limited for a subsequent period spanning more than a decade. Thereafter, research in MT picked up again, encouraged in no small part by the availability of computer systems with substantially enhanced processing power. Still, at that time, most of the MT systems developed involved what are termed as classic architectures (see Jurafsky and Martin 2009), which include:

(i)   direct translation systems, implementing word-by-word translations;
(ii) transfer approaches that first parse the input text and apply rules to transfer this parse structure to the target language;
(iii) interlingua approaches which transfer the input text from the SL to an abstract representation (termed interlingua) and from this generate the TL translation.

These architectures require a large proportion of human effort by specialists in order to create a working MT system.

Two main paradigms can be identified for MT systems. The first paradigm involves providing to the system explicitly the linguistic knowledge required to translate texts from the source to the target language. Most frequently, this is achieved by defining rules for this cross-lingual transfer, and thus the resulting methods are collectively known under the term Rule-based MT (RBMT).

As an alternative, it is possible to extract this linguistic knowledge from raw corpora, this paradigm being termed Corpus-based MT (termed as CBMT). The advantage of CBMT approaches lies in the hypothesis that language-specific information can be induced from appropriately collected and processed data rather than being handwritten explicitly as is the case in RBMT. Clearly, the latter paradigm requires a much more substantial computational effort to create a working MT system, which is why such methods have increased in popularity in the past two decades. On the other hand, CBMT provides the promise of substantially easier portability to new language pairs, and a reduced need for human input.

In more recent years, in a bid to achieve higher translation quality, researchers have attempted to merge principles from more than one MT paradigm, in order to combine their advantages. This effort has led to a family of methods collectively known as Hybrid MT (HMT). The different MT paradigms are briefly reviewed in the following subsection.

1.3 Advantages and Disadvantages of Main MT Paradigms

Rule-based Machine Translation is one of the oldest MT paradigms and still retains a measure of influence over modern MT systems. RBMT systems rely on compiling a comprehensive set of rules at various levels, e.g. in syntax, semantics, etc., for translating between two languages. RBMT systems have been developed for over 50 years and still remain one of the most popular paradigms because of their superior translation quality for specific languages, achieved by coding appropriately linguistic knowledge relevant to the source and target languages. Examples of RBMT systems include SYSTRAN, EUROTRA and more recently Apertium.

EUROTRA was initiated by the European Commission. The aim of this project was to create a small pre-industrial MT system capable of producing reasonably high-quality translations for written text between all official European Economic Community (EEC) languages of that time (seven at that time, cf. Arnold 1986). The set of languages was thereafter expanded to nine languages, to cover the newly added EEC countries.

EUROTRA adopted a stratificational approach, where for each language pair a sequence of analysis steps was performed to reach progressively a more abstract level. Following this, the transfer from the source to the target language was performed at the highest (i.e. most abstract) level, and then the synthesis steps were performed in reverse order in the TL, finally leading to the translation. In order to cover all possible language pairs, individual research teams were formed in nine countries in all, to handle the analysis and synthesis mechanisms for each country. The development effort lasted for the best part of a decade. The process has involved large teams of computational linguists working on theoretical and implementation issues. EUROTRA has been judged as being successful in terms of implementing research, but not in terms of creating a working system even in its final "industrial" phase (Arnold and Sadler 1992).

An RBMT system that has been available commercially is the SYSTRAN system, which has been developed over a number of language pairs, though not concurrently, while not all language pairs have reached the same level of accuracy. A key role in the SYSTRAN system is played by the dedicated rich bilingual lexicon, coupled with the appropriate taggers and parsers. Information on the SYSTRAN methodology to create an MT system is provided by Senellart et al. (2001). More recently, SYSTRAN's methodology has evolved so as to allow for the rapid development of new language pairs (Surcin et al. 2007), by making most of the software code reusable, even if it is not possible to match the rate of development of CBMT systems, as discussed in the next paragraphs. Finally, for new language pairs a hybrid methodology has been adopted, using the classic SYSTRAN RBMT system augmented by a set of statistical methods (Yang et al. 2009), further converging towards a hybrid MT system.

Apertium (Forcada et al. 2011) has been proposed as a methodology for developing MT systems in cases where the languages are close to one another. This methodology has been shown to be effective in comparison with hand-built systems. Evaluation results over a variety of related language pairs have been promising, though not as high as other RBMT methodologies. Apertium initially processes the input sentence by chunking the sequence of input tokens into fixed-length sequences corresponding to syntactic phrases and implementing translation of tokens based on a lexicon. These sequences are then processed to transform the syntax of the SL towards the TL. In order to define templates for Apertium-type methodologies, and thus improve their accuracy of translation, recently statistical-based methods have been proposed that do not require linguistic knowledge about the languages involved in the translation (Sánchez-Cartagena et al. 2015).

The main disadvantage of the RBMT paradigm as a whole is that a considerable amount of manual effort needs to be invested to create a working system. Furthermore, as a rule it is less than straightforward to use directly the modules of an existing RBMT system to create a translation system for other language pairs. In addition, progress in RBMT is hindered mainly by inadequate grammar resources for many (mostly under-resourced) languages and the absence of appropriate lexical resources and methods that would enable correct disambiguation and lexical choice.

As a response to these shortcomings, the CBMT paradigm was developed. In CBMT, linguistic rules denoting the syntactic and semantic preferences of words, as well as word order, constitute a large part of the implicit information extracted from large electronic corpora of texts. As a result, much of the linguistic knowledge is retrieved directly from the corpora, while the provision of explicit language-specific rules and tools is minimised. CBMT methods have been supported by the development of higher-performance computing systems in the 1990s, which have rendered realistic the processing of large collections of corpora in order to extract language modelling information, mainly of a statistical nature. The two major approaches within CBMT are Example-based MT and Statistical MT, which are briefly discussed below.

In terms of research activity, the SMT paradigm is the most important CBMT representative. SMT is based on being able to process large amounts of texts to extract correspondences. The principle of SMT was introduced by Brown et al. (1993), and is inspired by information theory principles. The idea is that a text is translated according to the probability distribution that a string in the target language represents the translation of a string in the source language. To calculate this probability, two types of model are used, (i) a translation model (representing the transfer from SL to TL) and (ii) a language model (indicating the probability of encountering a target language string in the given language). Following the introduction of the SMT paradigm, a large volume of research has been carried out by numerous research groups to build on the performance of SMT. The most important subsequent developments are discussed in Koehn (2010), including for instance factored models, discriminative training and syntax-based variants. A main benefit of SMT is that it is directly amenable to new language pairs using the same set of algorithms. One drawback of SMT systems is that they require appropriate training data in the form of parallel corpora for extracting the relevant translation models.

Thus, to develop an SMT system from a source language to a target language, SL-TL parallel corpora (where the same text is expressed in both the source and the target language) of the order of millions of tokens are required to allow the extraction of meaningful models for translation. Such corpora are hard to obtain, particularly for less resourced languages and are frequently restricted to a specific domain (or a narrow range of domains). In addition, it is accepted that existing parallel corpora are not suitable for creating general-purpose MT systems that focus on other domains. For this reason, in SMT, researchers are increasingly using syntax-based models as well as investigating the extraction of information from monolingual corpora, including lexical translation probabilities (Koehn and Knight 2002; Klementiev et al. 2012) and topic-specific information (Su et al. 2012).

The other key CBMT family is that of EBMT, where translations are generated via a reasoning-by-analogy process, as introduced conceptually by Nagao (1984). EBMT is based on having a large set of known pairs of sentences, each pair including one input sentence (in SL) and its corresponding translation (in TL). In EBMT, translations are generated by analogy, by appropriately processing the large library of examples during the actual translation phase rather than during a learning/training phase (Wu 2005), selecting the best matching one to the input sentence, and then replacing on the translation side the appropriate elements (for instance, tokens or phrases). A comprehensive review of the field of EBMT is provided by Hutchins (2005). The key difficulty in EBMT concerns searching through the entire set of sentence pairs to determine the one whose SL side best matches the input sentence, and then making the appropriate replacements.

In a bid to achieve higher translation quality, researchers have studied approaches that combine principles from more than one MT paradigm, in order to combine their advantages. This effort has led to a family of methods collectively known as Hybrid MT. Examples of HMT include the systems by Eisele et al. (2008) and Quirk and Menezes (2006), while the latest HMT activity is reported by Costa-jussà et al. (2013). The general convergence of more recent MT methodologies towards the combination of the most promising characteristics of each paradigm has been documented by Wu (2005), having started from pure MT systems belonging to one of the main paradigms (for instance, RBMT, SMT or EBMT) and increasingly progressing towards systems that combine characteristics from multiple paradigms. Thus, the effort to continuously improve translation quality leads to less clearly separable MT systems.

Concluding this brief survey, it is widely accepted that the most popular CBMT method is SMT, which needs parallel corpora aligned at a sentence basis. The popularity of SMT is closely related to the fact that open code has been released for use by researchers and MT practitioners, via the Moses software package (Koehn et al. 2007), and further developments in terms of open-source software are supported. This fact renders the creation of an MT system straightforward, provided that suitable resources are available. However, the compilation of parallel corpora is both expensive and time-consuming. Thus, alternative techniques have been studied for creating MT systems requiring resources which may be less informative but are also less expensive to collect or to create from scratch. These aim to eliminate the parallel corpus needed in SMT (or at least drastically reduce its size), employing instead either comparable corpora or monolingual corpora. Monolingual resources can be easily assembled for any language, for instance by harvesting the Web with relatively low effort. Methods following this approach had been proposed by Dologlou et al. (2003); Carbonell et al. (2006); Markantonatou et al. (2009). Though these methods do not provide a translation quality as high as SMT, their ability to develop MT systems with a very limited amount of specialised resources represents a valid starting point which is interesting in terms of research. This enables the development of a working MT system even for language pairs with very few language resources.

It is on the basis of the aforementioned works that the PRESEMT methodology has been developed. The design brief for PRESEMT has been to create a language-independent methodology that—with limited resources—can translate unconstrained texts giving a quality suitable for gisting purposes. In this methodology, the key design decision is to use a large monolingual corpus to extract most of the information required for the MT system. Thus, the dependence on a large parallel corpus is alleviated, requiring only a small parallel corpus (whose size is only a few hundred sentences) to provide information on the mapping of sentence structures from SL to TL. According to the preceding review of MT systems, PRESEMT can be classified as a Hybrid MT system, based on the argumentation of Quirk and Menezes (2006) and Wu (2005), combining certain elements of EBMT and SMT. The question, thus, becomes to what extent this methodology is more readily portable to new language pairs, while at the same time providing a translation accuracy which is comparable to that of state-of-the-art MT systems, and whether the translation quality is not too heavily compromised by the design decisions.

1.4 The PRESEMT Methodology in a Nutshell

To introduce the design principles of PRESEMT, the most intuitive way is to consider a potential application. An assumption is made that a person wishes to locate some piece of information over the Web. Most search engines retrieve Web pages which can cover the entire globe based on specific keywords, and naturally the corresponding text may be written in any language. As the use of the Internet expands, it is highly likely that the users will retrieve documents from several languages. Therefore, to obtain information, especially in the cases where only a few documents answering a query may be available, it is more than helpful to the users to provide an automatic translation of sufficient quality just for gisting. In that respect, the requirement is to transform the information, which is expressed in a given language, in the individual's native language.

A number of automatic translation systems are available over the Web, such as Google Translate, Bing Translator, SYSTRAN, where the user is prompted to either enter a text or alternatively define a Web page to receive their translation. As a rule, such MT systems currently give rather poor translations, apart from specific language pairs, which have been extensively developed and fine-tuned. In addition, it can be seen that for language pairs involving less widely-used languages subsequent versions of these MT systems sometimes even display a deterioration rather than improvement of the translation quality (examples of such non-monotonic behaviour will be discussed in subsequent chapters).

What is required from the user's point of view is a higher level of quality that is draft, but comprehensible as far as the average user is concerned. However, due to the natural language complexity it would probably be too complex a task to design a system that can produce translations of an appropriate quality for each and every domain. What is probably of more interest is to address specific user needs.

This necessitates developing an MT system that can be rapidly adapted to cover a new language pair, as well as to allow the user to effectively modify an existing MT system for a given language pair so that it better matches his/her requirements. The related set of requirements has formed the starting step for the PRESEMT project. This system needs to be characterised by the need for limited reliance on specialised resources, inherent language independence and the ability to modify the language resources used (e.g. the corpora used).

The main requirements for the PRESEMT system are to generate translations fast (a real-time or near real-time response is of prime importance) and to be able to develop new language pairs in a simple manner, without requiring specialised linguistic tools. In the modern multilingual environment of the European Union as well as beyond the boundaries of the Union itself, there exists an increased requirement for creating translation systems even for language pairs with limited availability of linguistic tools.

To cover these main requirements, the decision was taken to adopt cross-disciplinary ideas, mainly borrowed from the machine learning and computational intelligence domains. Thus, the core METIS idea (Dologlou et al. 2003) of combining machine-learning approaches with large monolingual corpora is retained in PRESEMT.

This idea is enhanced by a repertoire of pattern recognition and artificial intelligence techniques for linguistic applications ranging from the alignment of sentences and the creation of compatible phrases in different languages to the optimisation of system parameters. In this way, it is expected that substantial progress will be achieved in terms of (i) translation quality versus speed and (ii) language portability and ease of development of new language pairs. The two key objectives of the PRESEMT methodology are listed below:

Objective 1: Flexibility and adaptability—The MT system must be adaptable to user requirements and preferences, thus making it possible to address the issue of online translation for the masses. To that end, the PRESEMT MT system has a modular architecture, in order to maintain the integrity of the individual modules and facilitate local (i.e. module-specific) modifications, without affecting the system as a whole. For instance, the user may choose to retrain the MT system by retrieving corpora from the Web, having thus the freedom to specify resources, thematic domains and languages, to which the system will be called to adapt.

Objective 2: Language independence—The MT must be customisable to new languages. It is based on a language-independent method ensuring easy and cost-effective portability to new language pairs, without significant restrictions in the choice of either source or target languages. The support of the language-independent aspect of the prototype reduces significantly the human effort involved in the collection and processing (annotation, validation, etc.) of textual resources, as Web-sourced content serves as the major source of linguistic knowledge. At the same time, the provision of a list of key required resources (many of which can be shared between languages or easily modified when changing the language pair handled) supports the user in creating a new language pair within a limited number of days, by making use of existing modules and adapting resources accordingly.

1.5 Closing Note on Implementation

The issue of implementation and utilisation is of prime importance. Thus, the implemented MT system discussed in this volume is available for download and experimentation. The reader may visit the project's website to gain more information on the MT system and download the PRESEMT package coupled with some initial resources for the German-English and Greek-English language pairs that allow the running of an initial system. As an alternative, the fully functional online MT system can be accessed over the link www.presemt.eu/live. The website also provides detailed technical documentation and links to the standalone versions of the major PRESEMT modules hosted at Google Code, to encourage reuse.

References

Arnold D (1986) Eurotra: a European Perspective on MT. Proc IEEE 74(7):979–992

Arnold D, Sadler L (1992) EUROTRA: an assessment of the current state of the EC's MT programme. In: Translation and the computer, 13: the theory and practice of machine translation—a marriage of convenience? ASLIB, London

Brown PF, Della Pietra SA, Della Pietra VJ, Mercer RL (1993) The mathematics of statistical machine translation: parameter estimation. Comput Linguist 19(2):263–311

Carbonell J, Klein S, Miller D, Steinbaum M, Grassiany T, Frey J (2006) Context-based machine translation. In: Proceedings of the 7th AMTA Conference, Cambridge, MA, USA, pp 19–28

Costa-jussà MR, Banchs R, Rapp R, Lambert P, Eberle K, Babych B (2013) Workshop on hybrid approaches to translation: overview and developments. In: Proceedings of the 2nd HYTRA workshop, held within ACL-2013, Sofia, Bulgaria, pp 1–6

Dologlou Y, Markantonatou S, Tambouratzis G, Yannoutsou O, Fourla S, Ioannou N (2003) Using monolingual corpora for statistical machine translation: the METIS system. In: Proceedings of the EAMT-CLAW'03 Workshop, Dublin, Ireland, 15–17 May, pp 61–68

Eisele A, Federmann C, Uszkoreit H, Saint-Amand H, Kay M, Jellinghaus M, Hunsicker S, Herrmann T, Chen Y (2008) Hybrid machine translation architectures within and beyond the

EuroMatrix project. In: Hutchins J, v.Hahn W (eds) Proceedings of EAMT 2008 Conference, 22–23 September 2008, Hamburg, Germany, pp 27–34

Forcada ML, Ginestí-Rosell M, Nordfalk J, O'Regan J, Ortiz-Rojas S, Pérez-Ortiz JA, Sánchez-Martínez F, Ramírez-Sánchez G, Tyers FM (2011) Apertium: a free/open-source platform for rule-based machine translation. Mach Transl 25:127–144

Hutchins J (1996) ALPAC: the (in)famous report. MT News Int 14:9–12

Hutchins J (2005) Example-based machine translation: a review and commentary. Mach Transl 19:197–211

Jurafsky D, Martin JH (2009) Speech and language processing: an introduction to natural language processing. In: Computational linguistics and speech recognition, 2nd edn. Pearson Educational, Upper Saddle River, pp 895–944. ISBN 978-0-13-504196-1

Klementiev A, Irvine A, Callison-Burch C, Yarowsky D (2012) Toward statistical machine translation without parallel corpora. In: Proceedings of EACL2012, Avignon, France, 23–25 April, pp. 130–140

Koehn P (2010) Statistical machine translation. Cambridge University Press. xii, 433 pp. ISBN 978-0-521-87415-1

Koehn P, Knight K (2002) Learning a translation lexicon from monolingual corpora. In: Proceedings of the ACL-02 workshop on Unsupervised lexical acquisition, Philadelphia, Pennsylvania, U.S.A., 12 July 2002, pp. 9–16

Koehn P, Hoang H, Birch A, Callison-Burch C, Federico M, Bertoldi N, Cowan B, Shen W, Moran C, Zens R, Dyer C, Bojar O, Constantin A, Herbst E (2007) Moses: open source toolkit for statistical machine translation. In: ACL 2007: proceedings of demo and poster sessions, Prague, Czech Republic, June 2007, pp 177–180

Markantonatou S, Sofianopoulos S, Yannoutsou O, Vassiliou M (2009) Hybrid machine translation for low- and middle-density languages. In: Nirenburg S (ed) Language engineering for lesser-studied languages. IOS Press, pp 243–274. ISBN 978-1-58603-954-7

Nagao M (1984) A framework of a mechanical translation between Japanese and English by analogy principle. In: Elithorn A, Banerji R (eds) Artificial and human intelligence: edited review papers presented at the international NATO Symposium, October 1981, Lyons, France. Amsterdam: North Holland, pp 173–180

Quirk C, Menezes A (2006) Dependency treelet translation: the convergence of statistical and example-based machine translation? Mach Transl 20:45–66

Sánchez-Cartagena VM, Pérez-Ortiz JA, Sánchez-Martínez F (2015) A generalised alignment template formalism and its application to the inference of shallow-transfer machine translation rules from scarce bilingual corpora. Comput Speech Lang (Special Issue on Hybrid Machine Translation) 32(1):49–90

Senellart J, Dienes P, Varadi T (2001) New generation SYSTRAN system. In: Proceedings of the 8th MT Summit, 18–22 September, Santiago de Compostella, Spain, pp 311–316

Surcin S, Lange E, Senellart J (2007) Rapid development of new language pairs at SYSTRAN. In: Proceedings of MT Summit XI, 10–14 September, Copenhagen, Denmark, pp 443-449

Su J, Wu H, Wang H, Chen Y, Shi X, Dong H, Liu Q (2012) Translation Model Adaptation for Statistical Machine Translation with Monolingual Topic Information. In: Proceedings of ACL2012 Conference, Jeju, Republic of Korea, July 2012, pp 459–468

Wu D (2005) MT model space: statistical versus compositional versus example-based machine translation. Mach Transl 19:213–227

Yang J, Enoue S, Senellart J, Croiset T (2009) SYSTRAN Chinese-English and English-Chinese hybrid machine translation systems. In: Proceedings of CWMT-2009: the 5th China Workshop on Machine Translation, Nanjing, China, 16–17 October, p 8

Chapter 2
Implementation

This chapter introduces the general design characteristics of PRESEMT and provides a detailed description of all resources required as well as all pre-processing steps needed, such as corpora processing and model creation.

2.1   Introduction: Summary of the Approach

PRESEMT proposes a novel paradigm, which supports the straightforward development of MT systems for new language pairs using only a limited size of linguistic resources, by applying pattern recognition principles in a modular architecture. With respect to corpora, PRESEMT relies on two different sources of linguistic content, a large TL monolingual corpus and a small parallel corpus, typically comprising of a few hundred aligned SL-TL sentence pairs. Such resources can easily be collected from the Web, because one can easily find plenty of monolingual corpora for almost any language and the parallel corpus required is so small that it can even be assembled by hand. Therefore, PRESEMT overcomes one of the most important bottlenecks of all statistical systems (Munteanu and Marcu 2005): the availability of large parallel corpora of adequate quality. Such corpora are hard to find, particularly when not so widely used languages, such as Greek and Norwegian, are involved. The quality of the translation of such systems depends on the size and quality of the parallel corpus. Even if such corpora exist, they are frequently restricted to a very specific domain, such as the European Parliament (Koehn 2005). In addition to the size, the quality of the parallel corpus is also very important, but this is a factor that most MT research papers avoid to touch, as statistical methods promise that data size overcomes any minor quality issues that the data might have. But this is an important issue for parallel corpora that involve less popular languages, as size is comparably limited and quality-related issues usually increase as the text used is increased. In PRESEMT, both corpora are processed by their respective modules to produce the resources and models required

© The Author(s) 2017
G. Tambouratzis et al., *Machine Translation with Minimal Reliance on Parallel Resources*, SpringerBriefs in Statistics,
DOI 10.1007/978-3-319-63107-3_2

by the translation process. Finally, the modular architecture allows the language pair developer to replace one algorithm for another, as long as the strict requirements regarding data input and output formats are met.

Besides the two corpora described above, PRESEMT requires a bilingual lemma dictionary, a tagger-lemmatiser in both languages, and a shallow parser in the target language, that divides the text into non-overlapping sub-sentential segments (linguistic chunks). The system uses the TL parser to map this information to SL. In other words, given a parser (or more generally a phrasing model) in TL, one can generate an appropriate phrasing model in SL using pattern recognition-based clustering techniques. This is achieved in PRESEMT by using the parallel corpus to learn structural correspondences between the two languages in order to create sub-sentential segments which correspond to one another based on the structure of the parallel sentence. The modules implementing this functionality are the Phrase Aligner Module (PAM) and the Phrasing Model Generation (PMG).

Using the linguistic information provided by the shallow parser and the PAM + PMG combination, PRESEMT breaks down the translation process in two separate steps, the first one handling the order of both (i) chunks and (ii) out-of-chunk words in the final sentence, as well as the disambiguation of the more important tokens in the sentence, while the second step produces in parallel translations for all the chunks in the sentence. These two steps that breakdown the translation process in a divide-and-conquer fashion are implemented in the Structure Selection (SS) and the Translation Equivalent Selection (TES) modules, respectively, as discussed in Chap. 3. Figure 2.1 provides an overview of the PRESEMT translation process and the way that resources feed the modules. The following section describes all resources required to generate a working MT system via PRESEMT and all pre-processing steps needed.

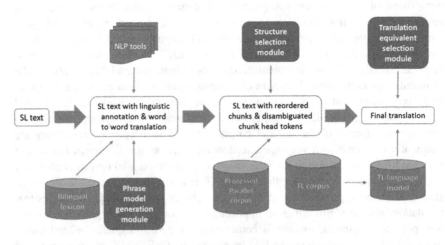

Fig. 2.1 Schematic representation of the translation process in PRESEMT

2.2 Linguistic Resources: Data and Existing Linguistic Tools

The system description starts by identifying the required linguistic resources and thus delineating a number of design decisions. A complete list of language-related information is provided, to indicate the basis upon which the specific MT system is created. This list comprises (i) a very small parallel corpus (a few hundred sentence pairs), (ii) a large monolingual corpus of the TL language (where arbitrarily large document collections can be handled) and (iii) a bilingual dictionary with lemmas. Additional linguistic information is provided in the form of Part-of-Speech (PoS) tagging and lemmatising for both languages, and a shallow parser for the TL language with phrase head information. Thus, only minimal resources are required for a prototype translation system to be created. A detailed record of all required linguistic resources follows.

2.2.1 External Processing Tools

The PRESEMT methodology relies on the application of linguistic annotation to all resources: corpora (monolingual and bilingual) and dictionaries, on the basis that resources of smaller size (in relation to those used by traditional SMT systems) may provide more useful information for using in an automatic translation task if linguistically annotated, and not just used to extract n-gram tables. The tools required are a tagger-lemmatiser for both languages and a shallow parser for only the target language.

Lemmatisation and Part-of-Speech tagging helps counterbalance data sparseness, which is a very important issue for PRESEMT because of the limited size of the resources used. Depending on the available tools as well as the language itself, part-of-speech tagging might contain morphological information such as case, number, gender and tense, all of which can greatly improve translation quality.

Syntactic phrases (chunks) give a glimpse of how sentences are formed by providing a flat structure annotation, with the main categories being noun and verb chunks and possibly with clause boundaries. While translating from one language to the other, chunking might also provide information about how groups of words might move from one side of the sentence to the other. Parsing for the source language is derived from the TL parsing scheme using the PMG module, because this way we tackle the problem of having completely different phrase segmentation between source and target languages.

All tools adopted for use in PRESEMT are pre-existing ones instead of tools being designed with the application in mind. Furthermore, these tools are not modified before their integration, so as to perform a realistic evaluation of the worthiness of the PRESEMT concept. This means of course that any errors produced by those tools propagate through the translation process. In order for a tagger-lemmatiser or a chunker to be used in the system, we must provide an interface for the system to interact with the existing tool, that being a simple script, installable software or a Web service. This interface is provided by building a java wrapper using as a guide one of the existing java wrappers created for tools such as the FBT tagger for Greek (Prokopidis et al. 2011) or the TreeTagger (Schmid 1994), which are included in the PRESEMT software package. In the specific case of shallow parsers, along with the java wrapper the developer must also build an accompanying resource for the identification of the head and functioning head (if available) for each phrase type, in the form of an XML file containing regular expressions. This information is essential, as head tokens provide crucial information in the whole translation pipeline. Figure 2.2 provides a sample of the HeadCriteria.xml file.

```
<headCriteria tool="TreeTagger">
<language init="EN">
<phrase type="pc" priority="right" fpriority="left">
    <head>^n.*</head>
    <head>^ex.*</head>
    <head>^fw.*</head>
    <head>^cd.*</head>
    <head>^jj.*</head>
    <head>^pp.*</head>
    <head>^wp.*</head>
    <head>^wd.*</head>
    <head>^pdt.*</head>
    <head>^dt.*</head>
    <head>^v.*</head>
    <fhead>in</fhead>
    <fhead>to</fhead>
</phrase>
<phrase type="nc" priority="right" fpriority="left">
    <head>^n.*</head>
    <head>^ex.*</head>
    <head>^fw.*</head>
    <head>^cd.*</head>
    <head>^jj.*</head>
    <head>^pp.*</head>
    <head>^wp.*</head>
    <head>^wd.*</head>
    <head>^pdt.*</head>
    <head>^dt.*</head>
```

Fig. 2.2 Head criteria file for the TreeTagger in English

2.2.2 Lemma-Based Bilingual Dictionary

The bilingual dictionary contains lemma forms of single-word and multi-word SL-TL lexical correspondences. In addition, it contains linguistic annotations, namely Part-of-Speech tags. The dictionary is a very important resource in the PRESEMT translation methodology and must provide a wide coverage of the source language to support a good translation quality. In theory, the larger the dictionary size, the fewer the out-of-vocabulary (OOV) words are expected to be. In PRESEMT, most of the dictionaries were based on ones provided by publishers and did not contain any linguistic annotations in appropriate and systematic ways, so a pre-processing step was necessary before using them in the MT system.

Table 2.1 provides the sizes of the dictionaries used during the PRESEMT project. As can be seen, different dictionary sizes have been adopted, depending on availability. For the system to be able to use a dictionary, this needs to be provided in the respective format. Figure 2.3 shows the XML representation used for storing the dictionaries in PRESEMT.

2.2.3 The Parallel Corpus

The parallel corpus used in PRESEMT needs to contain only a few hundred sentences, as it is only used for mapping the transfer from SL to TL sentence structures, determined as sequences of phrases. The small size of the corpus reduces reliance on costly linguistic resources. The corpus is assembled either from available parallel corpora or by using a Web crawler (Pomikálek and Rychlý 2008) and then manually replacing free translations with more literal ones, to allow the accurate extraction of structural modifications. After building the parallel corpus, we process the source and target side, using the SL and TL tagger-lemmatisers and the TL shallow parser, so as to annotate it with linguistic information. The result is a source and target side incorporating lemma and PoS information and other salient language-specific morphological features (e.g. case, number, tense, etc.) depending on the morphology of the language and the available tools. Furthermore, for the TL

Table 2.1 Dictionaries size for various language pairs used in PRESEMT

Language pair	Source	Number of entries
Greek-English	Developed in other project	40,000
Greek-German	Publisher	80,000
English-German	Developed in other project	1,000,000
Norwegian-English	Publisher	45,000
Norwegian-German	Publisher	37,000
Czech-English	Publisher	180,000
Czech-German	Publisher	70,000

```
<entry id="154">
    <slLemma tag="nncm">ακολουθία</slLemma>
    <tlLemma tag="nn">retinue</tlLemma>
</entry>
<entry id="155">
    <slLemma tag="nncm">ακολουθία</slLemma>
    <tlLemma tag="nn">service</tlLemma>
</entry>
<entry id="156">
    <slLemma tag="nncm">ακόλουθος</slLemma>
    <tlLemma tag="nn">attendant</tlLemma>
</entry>
<entry id="157">
    <slLemma tag="nncm">ακουστικό</slLemma>
    <tlLemma tag="nn">hearing</tlLemma>
    <tlLemma tag="nn">aid</tlLemma>
</entry>
<entry id="158">
    <slLemma tag="aj">ακουστικός</slLemma>
    <tlLemma tag="jj">auditory</tlLemma>
```

Fig. 2.3 Sample of the Greek-English dictionary

side, all sentences are split into non-overlapping syntactic phrases using a target language shallow parser. As the proposed methodology has been developed to maximise the use of publicly available software, the user is free to select any desired tools for these pre-processing tasks and there are no restrictions in using any available tool, as long as the developer completes the required integration tasks.

The parallel corpus is stored as two separate XML documents: one containing the tagged-lemmatised SL side and a second one with the tagged-lemmatised and chunked TL side. Samples of these documents can be seen in the extract of a Greek-English parallel corpus in Figs. 2.4 and 2.5. Notably, though the TL side of the corpus is split into phrases, the SL side consists of only the sentence words in sequence, without any information of the corresponding phrases. After the aforementioned preparation of the bilingual corpus, it is passed on to the Phrase Aligner Module for the identification of the corresponding words between SL and TL. The output of PAM in turn is passed to the Phrasing Model Generation module for the production of a parsing scheme on the source side, which will be used to process arbitrary sentences in SL and split them into the corresponding phrases.

```
<text>
  <sent id="1">
    <word id="1" head="n" fhead="n" token="H" tag="AtDfFeSgNm" lemma="o"/>
    <word id="2" head="n" fhead="n" token="Ευρωπαϊκή" tag="AjBaFeSgNm" lemma="ευρωπαϊκός"/>
    <word id="3" head="n" fhead="n" token="Ένωση" tag="NoCmFeSgNm" lemma="ένωση"/>
    <word id="4" head="n" fhead="n" token="δημιουργήθηκε" tag="VbMnIdPa03SgXxPePvXx" lemma="δημιουργώ"/>
    <word id="5" head="n" fhead="n" token="με" tag="AsPpSp" lemma="με"/>
    <word id="6" head="n" fhead="n" token="σκοπό" tag="NoCmMaSgAc" lemma="σκοπός"/>
    <word id="7" head="n" fhead="n" token="να" tag="PtSj" lemma="να"/>
    <word id="8" head="n" fhead="n" token="τερματιστούν" tag="VbMnIdXx03PlXxPePvXx" lemma="τερματίζω"/>
    <word id="9" head="n" fhead="n" token="οι" tag="AtDfMaPlNm" lemma="o"/>
    <word id="10" head="n" fhead="n" token="συχνοί" tag="AjBaMaPlNm" lemma="συχνός"/>
    <word id="11" head="n" fhead="n" token="και" tag="CjCo" lemma="και"/>
    <word id="12" head="n" fhead="n" token="αιματηροί" tag="AjBaMaPlNm" lemma="αιματηρός"/>
    <word id="13" head="n" fhead="n" token="πόλεμοι" tag="NoCmMaPlNm" lemma="πόλεμος"/>
    <word id="14" head="n" fhead="n" token="μεταξύ" tag="AdXxBa" lemma="μεταξύ"/>
    <word id="15" head="n" fhead="n" token="γειτονικών" tag="AjBaFePlGe" lemma="γειτονικός"/>
    <word id="16" head="n" fhead="n" token="χωρών" tag="NoCmFePlGe" lemma="χώρα"/>
    <word id="17" head="n" fhead="n" token="που" tag="PnReMa03PlNmXx" lemma="που"/>
    <word id="18" head="n" fhead="n" token="κορυφώθηκαν" tag="VbMnIdPa03PlXxPePvXx" lemma="κορυφώνω"/>
    <word id="19" head="n" fhead="n" token="με" tag="AsPpSp" lemma="με"/>
    <word id="20" head="n" fhead="n" token="το" tag="AtDfMaSgAc" lemma="o"/>
    <word id="21" head="n" fhead="n" token="Δεύτερο" tag="NmOdMaSgAcAj" lemma="δεύτερος"/>
    <word id="22" head="n" fhead="n" token="Παγκόσμιο" tag="AjBaMaSgAc" lemma="παγκόσμιος"/>
    <word id="23" head="n" fhead="n" token="Πόλεμο" tag="NoCmMaSgAc" lemma="πόλεμος"/>
    <word id="24" head="n" fhead="n" token="." tag="PTERM_P" lemma="."/>
  </sent>
</text>
```

Fig. 2.4 SL part sample of a Greek-English parallel corpus in PRESEMT

2.2.4 The TL Monolingual Corpus

The TL monolingual corpus is significantly larger than the parallel one and can be considered as the main resource for the main translation pipeline, as it is used to build the TL language model responsible for most of the translation tasks. The corpus size is of the order of a billion tokens. For example, the size of the English monolingual corpus used by the Greek-English PRESEMT system contains 3.65 billion tokens, while the size of the German one used by the Greek-German PRESEMT system contains 3.0 billion tokens. However, it should be stressed that the larger size of the monolingual corpus does not represent a constraint to the system creation as for most languages monolingual corpora of a good quality are available and most pre-processing is done offline, to speed up the translation process. All monolingual corpora created in PRESEMT were collected using a Web crawler (Pomikálek and Rychlý 2008). Before they can be used in the system, they are tagged-lemmatised and chunked offline during the pre-processing stage in order to produce the language model. The corpora with all the relevant annotation information are also stored using the PRESEMT XML representation shown in Fig. 2.5.

```
<text>
  <sent id="1">
    <clause id="1" type="">
      <phrase id="2" type="PC">
        <word id="3" head="n" fhead="y" token="-" tag="-" lemma="-"/>
        <word id="4" head="n" fhead="n" token="The" tag="DT" lemma="the"/>
        <word id="5" head="n" fhead="n" token="European" tag="NP" lemma="European"/>
        <word id="6" head="y" fhead="n" token="Union" tag="NP" lemma="Union"/>
      </phrase>
      <phrase id="7" type="VC">
        <word id="8" head="n" fhead="y" token="is" tag="VBZ" lemma="be"/>
        <word id="9" head="y" fhead="n" token="set" tag="VVN" lemma="set"/>
      </phrase>
      <phrase id="10" type="PRT">
        <word id="11" head="y" fhead="n" token="up" tag="RP" lemma="up"/>
      </phrase>
      <phrase id="12" type="PC">
        <word id="13" head="n" fhead="y" token="with" tag="IN" lemma="with"/>
        <word id="14" head="n" fhead="n" token="the" tag="DT" lemma="the"/>
        <word id="15" head="y" fhead="n" token="aim" tag="NN" lemma="aim"/>
      </phrase>
      <phrase id="16" type="PC">
        <word id="17" head="n" fhead="y" token="of" tag="IN" lemma="of"/>
        <word id="18" head="y" fhead="n" token="ending" tag="VVG" lemma="end"/>
      </phrase>
      <phrase id="19" type="PC">
        <word id="20" head="n" fhead="y" token="-" tag="-" lemma="-"/>
        <word id="21" head="n" fhead="n" token="the" tag="DT" lemma="the"/>
        <word id="22" head="n" fhead="n" token="frequent" tag="JJ" lemma="frequent"/>
        <word id="23" head="n" fhead="n" token="and" tag="CC" lemma="and"/>
        <word id="24" head="n" fhead="n" token="bloody" tag="JJ" lemma="bloody"/>
        <word id="25" head="y" fhead="n" token="wars" tag="NNS" lemma="war"/>
      </phrase>
      <phrase id="26" type="PC">
        <word id="27" head="n" fhead="y" token="between" tag="IN" lemma="between"/>
        <word id="28" head="y" fhead="n" token="neighbours" tag="NNS" lemma="neighbour"/>
      </phrase>
      <word id="29" head="n" fhead="n" token="," tag="," lemma=","/>
      <phrase id="30" type="PC">
        <word id="31" head="n" fhead="y" token="-" tag="-" lemma="-"/>
        <word id="32" head="y" fhead="n" token="which" tag="WDT" lemma="which"/>
      </phrase>
      <phrase id="33" type="VC">
        <word id="34" head="y" fhead="y" token="culminated" tag="VVD" lemma="culminate"/>
      </phrase>
      <phrase id="35" type="PC">
        <word id="36" head="n" fhead="y" token="in" tag="IN" lemma="in"/>
        <word id="37" head="n" fhead="n" token="the" tag="DT" lemma="the"/>
        <word id="38" head="n" fhead="n" token="Second" tag="NP" lemma="Second"/>
        <word id="39" head="n" fhead="n" token="World" tag="NP" lemma="World"/>
        <word id="40" head="y" fhead="n" token="War" tag="NP" lemma="War"/>
      </phrase>
      <word id="41" head="n" fhead="n" token="." tag="SENT" lemma="."/>
    </clause>
  </sent>
</text>
```

Fig. 2.5 TL part sample of a Greek-English parallel corpus in PRESEMT

2.3 Processing the Parallel Corpus

The role of the parallel corpus is to produce a phrase-mapping scheme between the SL and TL using suitably chosen lemma and PoS information in both language sides and shallow parsing only in the TL. By only using a chunker in the TL side, PRESEMT avoids the use of an additional external tool, thus increasing portability to new language pairs, while also avoiding potential incompatibilities when creating alignments between words and phrases of the two languages.

The processing is performed in two stages. In the first stage, the TL side parsing scheme is transferred in the SL side by building word and phrase alignments using PAM. PAM transfers the TL side parsing scheme, which encompasses lemma, tag and chunking information (namely phrase boundaries and phrase labels), to the SL side, based on lexical information (retrieved from the lexicon) coupled with statistical data on PoS tag correspondences extracted from the lexicon. PAM follows a three-step process, defining alignments based on (a) lexicon entries, (b) similarity of grammatical features and PoS tag correspondence and (c) the alignments of neighbours of the unaligned words.

In the second stage, an SL phrasing model is constructed by PMG, by applying probabilistic methodologies to the PAM output. This phrasing model is then applied for segmenting any arbitrary SL text being input for translation. Initially, PMG was implemented using Conditional Random Fields (CRF), due to the high representational capabilities of this probabilistic model (Lafferty et al. 2001). Alternative approaches for building PMG based on template-matching principles have been investigated (cf. Tambouratzis et al. 2013), though unless otherwise stated the results reported in this volume utilise the CRF model, which is the default tool used within the PRESEMT methodology.

The following two sections provide a detailed description of (i) the Phrase Aligner and (ii) the Phrasing Model Generation modules, respectively.

2.3.1 Phrase Aligner Module

To determine a model expressing the transfer of phrases from SL to TL, it is essential to have the sentences of the parallel corpus analysed into pairs of corresponding phrases in SL and TL. Development work in the earlier MT system METIS-II (Markantonatou et al. 2009) has demonstrated that when trying to harmonise the phrasings from two independently created parsers/chunkers (where one operates in SL and one in TL), extensive effort is required for their modification to compatible phrasing schemes for SL and TL. Thus, such an approach is not suitable for a methodology intended to be ported to new language pairs with minimal effort. To that end, in PRESEMT the Phrase Aligner Module (Tambouratzis et al. 2011) is developed, to eliminate the need for an SL side parser. PAM is dedicated to transferring to the SL side the TL side parsing scheme, which encompasses phrase boundaries and phrase types. During this transfer, the TL side phrase type is inherited by the corresponding SL phrase. In terms of its two-phase approach, PAM has conceptual similarities to a number of works (cf. Och and Ney 2004; Ganchev et al. 2009), as initially (i) words in the SL sentence are aligned to those of the TL sentence and afterwards (ii) unaligned SL words are grouped into phrases depending on agreement of grammatical features.

PAM follows a process divided into three steps, where in each step the aim is to resolve the alignments for tokens that remain unaligned from earlier steps. In the first step, alignments are performed on the basis of the lemmas included in the bilingual lexicon. Thus, tokens between SL and TL are aligned if the lexicon indicates an equivalence in meanings (i.e. one is a valid translation of the other), provided that there are not multiple ambiguous matches between lexicon entries and SL or TL tokens. Later steps use more general information (such as the PoS tag of the tokens rather than lexical information) to align tokens and thus have a lower likelihood of producing the correct alignment than the first step. In the second step, alignments are determined based on the similarity of grammatical features between adjoining tokens (in morphologically rich languages the information of gender or case agreement may associate to one another related tokens for grouping in the same phrase). Finally, in the third step, unaligned words are aligned based on string similarity as well as on the alignments of their neighbours (under the assumption of locality of alignments). The entire alignment process is described in detail in Tambouratzis et al. (2012).

After the alignments on a token level are completed, the aim becomes to map for each phrase in TL all corresponding tokens in SL so as to create phrases in the SL side of the parallel corpus. This process results in the establishment of the SL side phrases, for each of which a correspondence to a TL phrase is defined.

To establish the SL side phrasing, PAM operates on the parallel corpus by utilising the following resources:

(1) a bilingual lexicon from SL to TL;
(2) an SL tagger-lemmatiser (which may provide both basic PoS characterisation and more refined features, i.e. case, number, person, etc.);
(3) a TL tagger-lemmatiser and shallow parser (which can again provide basic and refined features;
(4) a TL clause boundary detection tool.

Based on this set of inputs, PAM decides on the optimal segmentation of the source sentence into phrases. A multicriterion-type comparison is implemented, where the aforementioned inputs are prioritised and combined. Though not all aforementioned inputs need to be present for PAM to work, their use results in a more accurate alignment.

Alignment Step 1: Lexical information

The bilingual lexicon provides information on likely word and lemma correspondences between SL and TL. The word aligner algorithm performs alignment of SL words to TL words via the bilingual lexicon. The algorithm allows the one-to-one alignment between SL words and TL ones, while rejecting any multiple alignments, unless the lexicon explicitly provides such information. In the case of multiple possible alignments, the principle is that for every word k in SL that is potentially aligned to more than one word in TL, the TL word chosen is the one (a) that has the minimum distance from the single-aligned TL word and (b) for which the corresponding single-aligned SL word has the minimum distance in tokens from word k.

Correspondences of SL-TL PoS tags are also extracted, by running through the bilingual lexicon to estimate the likelihood of alignment between PoS tags, for instance to extrapolate if a verb in SL is more likely to translate to a verb or to a noun in TL. Such correspondences are used to determine alignments for out-of-vocabulary cases, where the lexicon does not provide sufficient information.

As an example of the PAM operation, let us consider the following pair of sentences:

(*Greek—SL*): *Με την Ευρωπαϊκή Κοινότητα Άνθρακα και χάλυβα αρχίζουν να ενώνονται οι Ευρωπαϊκές χώρες οικονομικά και πολιτικά.*
(*English—TL*): *The European Coal and Steel Community begins to unite European Countries economically and politically.*

It is assumed that the lexicon entries relevant to this pair of sentences are those listed in Table 2.2. Then, in Fig. 2.6, the SL and TL sides are depicted in the form

Table 2.2 Indicative lexicon entries in lemma and PoS tag form

Greek language		English language	
Lexicon entry	PoS tag	Lexicon entry	PoS tag
ευρωπαϊκός	Aj	European	JJ
άνθρακας	No	Coal	NN
χώρα	No	Country	NN

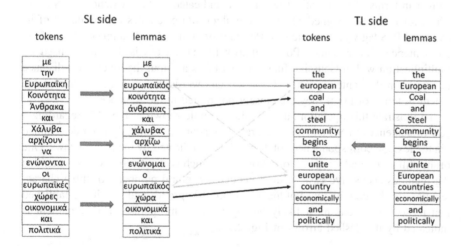

Fig. 2.6 Alignment of words in SL and TL based on the lexical information of Table 2.2

of token and lemma sequences in the left- and right-hand sides, respectively. The alignments that can be identified by the lexicon entries are depicted by arrows from SL to TL, where the dark arrows correspond to unambiguous alignments, while grey arrows indicate more than one possible alignment per SL and TL token.

Based on the entries of Table 2.2, there are unique correspondences in SL and TL for the pairs "άνθρακας"—"coal" and "χώρα"—"country". On the contrary, for the lexicon pair "Ευρωπαϊκός—European", two occurrences exist in both SL and TL and thus two pairs of possible alignments are in consideration (as noted before both SL tokens aligning to the same TL one—or vice versa—are not allowed). In this case, in the absence of any other knowledge, neighbouring tokens for which alignments are already established help to determine the most likely alignments. Thus, the alignments for "coal" and "country" are the preferred ones for the two instances of "European", as indicated by the solid grey arrows in Fig. 2.6, while the less likely alignments are indicated by dashed grey lines in the same figure.

When an SL word remains unaligned, usually due to limited dictionary coverage, the algorithm transliterates it (in case of different SL and TL alphabets, e.g. Greek and English) and consequently attempts to match it to a word with high similarity in the TL sentence. Two words, for which no association is indicated by the lexicon, are considered similar when their letter-wise similarity, in terms of the longest common sub-sequence ratio, exceeds a threshold.

At the end of Step 1, all possible alignments using lexical information have been established. SL words that remain unaligned are handled by subsequent steps using other types of information.

Alignment Steps 2 and 3: Similarity of features

Operating on the output of Step 1, subsequent steps attempt to assign unaligned SL tokens into phrases, by identifying nearby SL tokens which are aligned, and that are similar in terms of grammatical features (as indicated by their extended PoS tags). Thus, for every unassigned SL word the algorithm calculates the similarity of its extended PoS tag with the extended PoS tags of all the already aligned SL words in the sentence. The extended PoS similarity for each word is then normalised by multiplication with a Gaussian function that takes as input the token-wise distance of words on the sentence. Then, PAM clusters words that match to an acceptable extent in terms of tag if they are closely positioned in the sentence.

To illustrate this operation, assume the example shown in Fig. 2.7, regarding a pair of sentences in SL and TL. Several alignments have already been determined via the lexicon correspondences of Table 2.3, as indicated by the black arrows. One of the still unaligned words is "οικονομικά", which is an adjective whose extended tag includes accusative case, neutral gender and plural number. All three characteristics are shared with those of the token "προβλήματα", and thus it is established that these two tokens most likely belong to the same phrase. This alignment is indicated by the dashed arrows of Fig. 2.7.

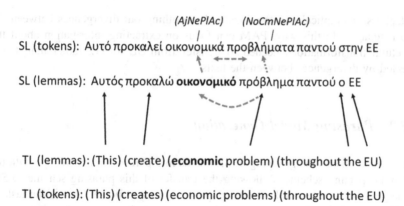

Fig. 2.7 Example of alignment making use of extended features of neighbouring words (tags are indicated in *brackets* for selected words)

Table 2.3 Lexicon entries relevant to the example of Fig. 2.7

Greek language		English language	
Lexicon entry	PoS tag	Lexicon entry	PoS tag
αυτός	Pn	This	DT
πρόβλημα	No	Problem	NN
ο	At	The	DT
εε	ABBR	Eu	NN

The final alignment of words to this sentence pair is indicated in Fig. 2.8, where the last pending alignments have been resolved. This is achieved by the high likelihood of correspondences within the lexicon entries between tokens with PoS tag "Av" in Greek and tokens with tag "ADV" in English (both of these corresponding to adverbs).

Regarding the bilingual corpora used, it is advisable to edit them so that the SL and TL sides of the corpus are as "close" as possible to each other, removing

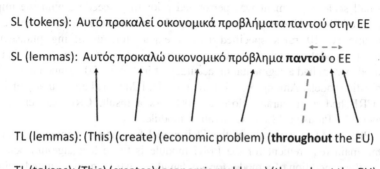

Fig. 2.8 Alignment based on the likelihood of PoS correspondences between SL and TL

metaphors or elliptical constructions and smoothing out divergences between the two languages. In this way, PAM can focus on extracting information about the structural transformations needed to transfer from SL to TL, rather than being affected by divergences between the texts.

2.3.2 Phrasing Model Generation

PAM accomplishes the grouping of SL tokens into phrases, in accordance with the given TL parsing scheme. Following the transfer of this phrasing scheme to SL, archetypes become available for developing a phrasing model. This is the task of the Phrasing Model Generation, which learns by example to segment arbitrary input text into phrases in compliance with the TL phrasing scheme. If this is achieved, the aligned parallel corpus can be used to transform the structure of input sentences from SL to TL.

When initiating the work on the Phrasing Model Generation, a survey of relevant work was undertaken to identify appropriate methods. Since this is a widely studied topic, it was decided to select the most promising existing technique (preferably one which employs free-to-use or open-source software), rather than developing or reimplementing a new technique. This can speed up the system development and makes use of already proven techniques (alternative model generation techniques are investigated in Chap. 6). Most relevant studies have converged at using a probabilistic methodology. It is widely accepted that among the statistical-based models used, Conditional Random Fields (Lafferty et al. 2001) provide the most promising avenue for creating parsers (e.g. Sha and Pereira 2003; Tsuruoka et al. 2009). Due to the expressiveness of the underlying mathematical model, CRF requires a large number of training patterns to extract an accurate model. In comparison with other probabilistic modes, CRF has been found to possess a superior performance to both Hidden Markov Models (HMMs) and Maximum Entropy (ME) models by avoiding biasing solutions towards states with few successor states (Wallach 2004).

A small scale experiment was performed prior to proceeding with the implementation of PMG. This experiment compared a custom rule-based system (comprising approx. 10 rules specified by an expert for identifying phrases) to probabilistic models, which was refined over three iterations. It was found that the rule-based system had a segmentation accuracy of just under 70%, much lower than probabilistic models. Among probabilistic models, HMM had an accuracy of 78%, while CRF had an accuracy close to 90%. As a result, CRF was chosen to implement the Phrasing Model Generation module.

The PRESEMT system utilises the CRF model for phrasal segmentation in the SL. One main requirement for the PMG module is to be language-independent, allowing the generation of a model for any language, working on a limited training set. As PRESEMT assumes a parallel corpus of at most a few hundred sentences, the model should be established taking into account the size of the training set.

Thus, one needs to move to a higher level of abstraction, beyond token (or lemma) forms.

Regarding the PMG algorithmic part, the MALLET package (McCallum 2002) was chosen as it is implemented in java, which is also used for the PRESEMT prototype. Different system set-ups were experimentally tested for the CRF model via the options available within MALLET.

Both the default CRF training method (hereafter denoted as "std.") and the alternative method (denoted as "alt.") were tested. Both the complete and reduced tagsets (denoted as "std." and "red.", respectively) were investigated, to determine the optimal configuration. Another consideration involves the detail used in the sequence of tokens. For each token in the SL side, there exist different levels of representation, at token, lemma and tag information. However, a training set of typically 200 sentences is definitely too small to extract a phrasing model based on lemmas, so it was decided to employ PoS tags to identify phrase boundaries.

Preliminary experimentation showed improved segmentation accuracy when using only the basic Part-of-Speech tag (such as "Vb" for verb or "No" for noun), coupled with the case for tokens that have this type of information. The model order was also varied, this taking into consideration only the information of the present word (model 0) but also the previous one (denoted as model 0-1) or even the two previous ones (model 0-1-2).

Indicative experimental results are reported in Table 2.4, where for each token, the PMG-generated phrasing information is compared to the gold-standard created by a language specialist over a 100-sentence development set. The segmentation accuracy is expressed as the fraction of tokens that are correctly assigned to their corresponding phrases. The CRF phrasing accuracy peaks at 90%, for the reduced tagset including only the PoS tag and the case feature. The best results were achieved when adopting a CRF model with 0-1 order, while higher model orders resulted in no measurable improvement. Hence, a second-order CRF is used as the default parser for the SL side in PRESEMT, using the last two tokens. It should be noted that the MALLET functionalities are integrated within PRESEMT as a separate module. Thus, the user may invoke via two commands the process that creates a new phrasing model, first performing alignment (via PAM) and then generating the phrasing model via the training of CRF.

Table 2.4 PMG experimental accuracies (denoted in percentages)

Feature	Parameters		Model order		
	Tags	Method	0	0-1	0-1-2
1-gram	Std.	Std.	75.4	80.4	77.8
1-gram	Red.	Std.	82.4	88.1	84.4
1-gram	Red.	Alt.	81.3	89.0	86.0
2-gram	Std.	Std.	73.5	74.8	73.3
2-gram	Red.	Std.	85.5	86.7	84.5
2-gram	Red.	Alt.	89.3	90.0	88.7

2.4 Creating a Language Model for the Target Language

The annotated TL corpus is used for the creation of a language model based on syntactic phrases. This model is then applied to the translation pipeline for establishing correct ordering of words within each phrase, disambiguating between alternative translations and handling functional words (for instance, insertion or deletion of articles or negation particles).

Unlike the statistical language models that are based on n-grams of words, the words here are grouped together based on the syntactic phrases extracted from the chunked TL monolingual corpus. All TL phrases are organised in a hash map, using as a key the following three factors: (i) phrase type, (ii) lemma of the phrase head and (iii) PoS tag of the phrase head. Each TL phrase extracted from the corpus is stored in the equivalent hash map along with its number of occurrences in the corpus. Finally, each map is serialised and stored in a separate file in the file system, with an appropriate name for easy retrieval, so that the system will not have to load the whole model in memory during run-time. For instance, for the English monolingual corpus, all verb phrases with the lemma of the head token being "read" and the PoS tag "VV" are stored in the file "Corpora/EN/Phrases/VC/read_VV".

As an example, let us assume a very small TL corpus consisting only of the following sentence: "*A typical scheme would have eight electrodes penetrating human brain tissue; wireless electrodes would be much more practical and could be conformal to several different areas of the brain*". The syntactic phrases extracted from this small corpus are shown in Table 2.5, while the files created for the model are shown in Fig. 2.9. Because all phrases only appear once in the corpus, the frequencies are omitted in the specific example.

It should be noted that, with respect to large corpora, in order to reduce the number of files created, if a TL phrase file remains very small (based on the definition of a small threshold value), i.e. it contains very few frequent phrases (less

Table 2.5 Syntactic phrases extracted from the TL monolingual corpus

ID	Phrase type	Phrase content	Phrase head lemma/PoS
1	PC	A typical scheme	Scheme/NN
2	VC	Would have	Have/VH
3	PC	Eight electrodes	Electrode/NN
4	VC	Penetrating	Penetrate/VV
5	PC	Human brain tissue	Tissue/NN
6	PC	Wireless electrodes	Electrode/NN
7	VC	Would be	Is/VB
8	PC	Much more practical	Practical/JJ
9	VC	Could be	Is/VB
10	PC	Conformal	Conformal/JJ
11	PC	To several different areas	Area/NN
12	PC	Of the brain	Brain/NN

File 1: VC/Have_VH	
2	Would have

File 2: VC/Is_VB	
7	Would be
9	Could be

File 3: VC/penetrate_VV	
2	penetrating

File 4: PC/scheme_NN	
1	A typical scheme

File 5: PC/electrode_NN	
3	Eight electrodes
6	Wireless electrodes

File 6: PC/Tissue_NN	
5	Human brain tissue

File 7: PC/Practical_JJ	
8	Much more practical

File 8: PC/conformal_JJ	
10	conformal

File 9: PC/areas_NN	
11	To several different areas

File 10: PC/brain_NN	
12	Of the brain

Fig. 2.9 Example of monolingual corpus phrases splits into files

Table 2.6 Statistics for the English and German monolingual corpora

	English	German
Size in tokens	3,658,726,327	3,076,812,674
Number of raw text files (*each cont. a block of ca. 1 Mbyte*)	87,000	96,000
Sentence number	1.0×10^8	9.5×10^7
Phrase number	8.0×10^8	6.0×10^8
Number of extracted phrase files	380,000	478,000

than 10 unique ones), it is not stored separately, but its content is moved in a file with similarly rare phrases.

Table 2.6 provides statistics for the phrase TL language models created for the English and German languages.

References

Ganchev K, Gillenwater J, Taskar B (2009) Dependency grammar induction via bitext projection constraints. In: Proceedings of the 47th Annual Meeting of the ACL, Singapore, 2–7 August, pp 369–377

Koehn P (2005) Europarl: a parallel corpus for statistical machine translation. MT Summit 2005, Phuket, Thailand

Lafferty J, McCallum A, Pereira F (2001) Conditional random fields: probabilistic models for segmenting and labelling sequence data. In: 28th International Conference on Machine Learning, ICML 2011, Bellevue, Washington, USA, pp 282–289

Markantonatou S, Sofianopoulos S, Giannoutsou O, Vassiliou M (2009) Hybrid machine translation for low- and middle-density languages. In: Nirenburg S (ed) Language engineering for lesser-studied languages. IOS Press, pp 243–274

McCallum AK (2002) MALLET: a machine learning for language toolkit. http://mallet.cs.umass.edu

Munteanu DS, Marcu D (2005) Improving machine translation performance by exploiting non-parallel corpora. Comput Linguist 31(4):477–504

Och FJ, Ney H (2004) The alignment template approach to statistical machine translation. Comput Linguist 30(4):417–449

Pomikálek J, Rychlý P (2008) Detecting co-derivative documents in large text collections. In: Proceedings of LREC2008, Marrakech, Morrocco, pp 1884–1887

Prokopidis P, Georgantopoulos B, Papageorgiou H (2011) A suite of NLP tools for Greek. In: Proceedings of the 10th ICGL Conference, Komotini, Greece pp 373–383

Schmid H (1994) Probabilistic part-of-speech tagging using decision trees. In: Proceedings of International Conference on New Methods in Language Processing, Manchester, UK, pp 44–49

Sha F, Pereira FCN (2003) Shallow parsing with conditional random fields. In: Proceedings of HLT-NAACL Conference, pp 213–220

Tambouratzis G, Simistira F, Sofianopoulos S, Tsimboukakis N, Vassiliou M (2011) A resource-light phrase scheme for language-portable MT. In: Proceedings of the 15th International Conference of the European Association for Machine Translation, 30–31 May, Leuven, Belgium, pp 185–192

Tambouratzis G, Troullinos M, Sofianopoulos S, Vassiliou M (2012) Accurate phrase alignment in a bilingual corpus for EBMT systems. In: Proceedings of the 5th BUCC Workshop, held within the LREC-2012 Conference, May 26, Istanbul, Turkey, pp 104–111

Tambouratzis G, Sofianopoulos S, Vassiliou M (2013) Language-independent hybrid MT with PRESEMT. In: Proceedings of HYTRA-2013 Workshop, held within the ACL-2013 Conference, Sofia, Bulgaria, 8 August, pp 123–130

Tsuruoka Y, Tsujii J, Ananiadou S (2009) Fast full parsing by linear-chain conditional random fields. In: Proceedings of the 12th EACL Conference, 30 March–3 April, Athens, Greece, pp 790–798

Wallach HM (2004) Conditional random fields: an introduction. University of Pennsylvania CIS Technical Report, MS-CIS-04-21, February 24

Chapter 3
Main Translation Process

3.1 Introduction

This chapter presents in detail the main translation process of PRESEMT, delving deeper in the core of the system and its inner workings.

All SMT systems (Koehn 2010), which represent the dominant approach to automatic translation, employ two statistical models in order to find the best translation in the target language: the translation model and the language model. The translation model derives from parallel corpora and can have many forms and employ various language features. Its task is to produce various TL alternative translations based on n-grams found in the SL input sentence and rank them according to their probabilities. The language model derives from the monolingual TL corpus and given a TL sentence its task is to provide a metric as to how well formed the TL is. By combining both models during the decoding process, an SMT system can provide translation alternatives using the translation model, and then reorder the words and make translation choices over the words of each translation alternative, finally selecting the translation with the highest score.

Local and long-distance reordering is one of the most challenging aspects of any Machine Translation system. In modern SMT, numerous approaches have used pre-processing techniques that perform word reordering in the source side based on the syntactic properties of the target side (Rottmann and Vogel 2007; Popovic and Ney 2006; Collins et al. 2005), in order to overcome the long-distance word reordering problem. Of course, short-range reorderings are easily captured by the language model if missed by the translation model.

PRESEMT differentiates itself from all modern SMT systems, by using a bilingual dictionary and breaking down the translation process in two steps, preceded by a pre-processing step (cf. Chap. 2). During pre-processing, the input SL text is tagged, lemmatised and chunked, using the phrasing model produced by the PMG. In the first step of the translation process (Structure Selection), the dictionary produces translation alternatives (single and multi-word ones) for all words in the

© The Author(s) 2017
G. Tambouratzis et al., *Machine Translation with Minimal Reliance on Parallel Resources*, SpringerBriefs in Statistics,
DOI 10.1007/978-3-319-63107-3_3

input text, while the parallel corpus provides a phrasing scheme for chunking the input text into linguistic segments and afterwards reordering them according to TL phrase patterns, on a sentence-by-sentence basis. In the second step (Translation Equivalent Selection), the language model extracted from the TL corpus is applied to each phrase in order to resolve lexical ambiguities, fix the word order and handle functional words such as articles and prepositions. If asked to classify PRESEMT into an MT paradigm, it would be put under hybrid systems, as the Structure Selection phase is closely related to EBMT (Nagao 1984; Hutchins 2005), while the second phase is reliant upon statistical information (Brown et al. 1988).

By breaking down the translation pipeline, PRESEMT deals with long-range and short-range reordering separately. The first step performs structural transformation of the source side in accordance with the syntactic phrases of the target side, trying to capture long-range reordering, while the second step makes lexical choices and performs local word reordering within each phrase. Because of the modular nature of PRESEMT, both translation pipeline modules can be replaced with alternatives, as long as they comply with the input and output format of the data. PRESEMT is a language-independent process, and minor changes in the translation modules may be needed when moving to new language pairs involving different language families, but the main algorithms will remain unchanged. Next, the two steps of the translation pipeline are described.

3.2   Translation Phase One: Structure Selection

After the pre-processing of the input text, it is passed on to the Structure Selection module to transform its structure sentence-by-sentence using the limited bilingual corpus as a structural knowledge base, closely resembling the "translation by analogy" aspect of EBMT systems (Nagao 1984). Using available structural information, namely the type of syntactic phrases, the Part-of-Speech tag of the head of each phrase and the case of the head (if available), we retrieve the most similar source side sentence from the parallel corpus.

As a schematic example, let us assume the parallel corpus illustrated in Fig. 3.1 (note that SL side phrases are depicted in capital characters, while TL side phrases are depicted in small letters and that the alignments between phrases are indicated with dashed lines).

Using the stored alignment information from the bilingual corpus between the source and target side, we must perform all necessary actions in order to transform the structure of each input sentence to the structure of the target side of the corpus sentence pair. Figure 3.2 provides a simplified overview of the comparison process, where an input sentence consisting of five phrases (more specifically the sequence "ADVC-PC-VC-PC-PC") is compared in terms of structure (sequence of phrases) to the four SL sentences from the parallel corpus (as depicted in Fig. 3.1). In this simple example, the fourth sentence of the parallel corpus has exactly the same structure to that of the input sentence. Thus, the structure of the translation should be "advc-pc-pc-vc-pc", as suggested by the TL side of the fourth sentence of the parallel corpus.

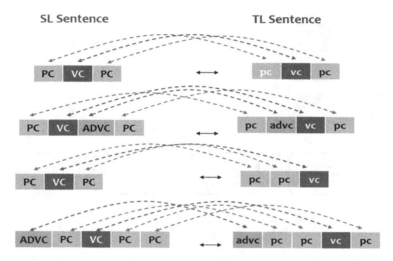

Fig. 3.1 SL-TL phrase alignments between sentence pairs in the parallel corpus (tags are omitted for simplicity reasons)

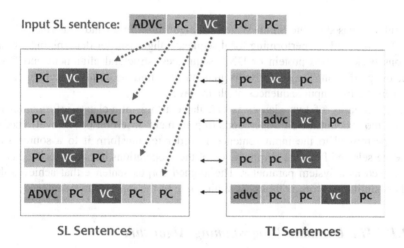

Fig. 3.2 Structural comparison of the input sentence to sentences on the SL side of the parallel corpus

Figure 3.3 depicts the processing steps performed by the Structure Selection module for a single SL sentence. Naturally, in the common case of multiple input sentences, the process is executed for each sentence separately (and in parallel using multiple threads), and passed on to the next phase of the translation pipeline.

For the retrieval of the most similar SL sentence from the parallel corpus, a dynamic programming algorithm was selected. Therefore, the Structure Selection process is treated as a sequence alignment, aligning the input sentence to an SL sentence from the parallel corpus and assigning a similarity score. The implemented

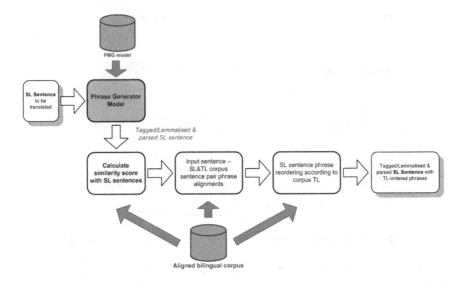

Fig. 3.3 Data flow in the Structure Selection module

algorithm is based on the Smith-Waterman algorithm (Smith and Waterman 1981), initially proposed for performing local sequence alignment for determining similar regions between two protein or DNA sequences, structural alignment and RNA structure prediction. The algorithm is guaranteed to find the optimal local alignment between the two input sequences at clause level.

The similarity of two clauses is calculated using intra-clause information by taking into account the edit operations (replacement, insertion or removal) needed to be performed to the input sentence in order to transform it to a source side sentence selected from the corpus. Each of these operations has an associated cost, considered as a system parameter. The aligned corpus sentence that achieves the highest similarity score is the most similar one to the input source sentence.

3.2.1 The Dynamic Programming Algorithm

The source sentence is parsed in accordance with the phrasing model created by the PMG module. The first step of the algorithm is to compare the input sentence (IS) to all the SL sentences of the parallel corpus in terms of structure. A two-dimensional table is built with each of the IS phrases occupying one column (the corresponding phrases being shown at the top of the table) and the candidate corpus sentence (CS) phrases each occupying one row (the corresponding CS phrases being shown along the left side of the table). A cell (i, j) represents the cumulative similarity of the sub-sequence of elements up to the mapping of elements E_i of CS and E'_j of IS. Elements refer to syntactic phrases, represented by their type and the PoS tag and case (where available) of each phrase head.

The value of cell (i, j) is filled by taking into account the cells directly to the left $(i, j - 1)$, directly above $(i - 1, j)$ and diagonally above-left $(i - 1, j - 1)$, these containing values $V1$, $V2$ and $V3$, respectively, and is calculated as the sum of the maximum of the three values plus the calculated similarity between the elements Ei and E'_j. While calculating the value of each cell, the algorithm also keeps tracking information of the chosen maximum value so as to allow the construction of the actual alignment vector.

The similarity of two phrases (**PhrSim**) is calculated as the weighted sum of the phrase type similarity (**PhrTypSim**), the phrase head PoS tag similarity (denoted as **PhrHPosSim**), the phrase head case similarity (**PhrHCasSim**) and the functional phrase head PoS tag similarity (**PhrfHPosSim**):

$$PhrSim(E_i, E'_j) = W_{phraseType} * PhrTypSim\left(E_i, E'_j\right) + W_{headPoS} * PhrHPosSim\left(E_i, E'_j\right)$$
$$+ W_{headCase} * PhrHCasSim\left(E_i, E'_j\right) + W_{fheadPoS} * PhrfHPosSim\left(E_i, E'_j\right)$$
$$(2.1)$$

In the default implementation of the algorithm, the weights have been initialised with specific values, their sum being equal to 1 for normalisation purposes, yet the optimal values are determined during an optimisation phase, which is an offline process executed independently from the Main Translation process.[1]

The similarity score ranges from **100** to **0**, the two extremes denoting exact match and total dissimilarity between two elements E_i and E'_j, respectively. In case of a zero similarity score, a penalty weight (-50) is employed, to further discourage selection of such correspondences.

When the algorithm has reached the jth element of the IS, the similarity score between the two SL clauses is calculated as the value of the maximum-scoring jth cell. The CS that achieves the highest similarity score is the closest to the input SL clause in terms of phrase structure information.

Apart from the final similarity score, the comparison table of the algorithm is used for finding the actual alignment of phrases between the two SL clauses. By combining the SL clause alignment from the algorithm with the alignment information between the CS and the attached TL sentence, the IS phrases are reordered according to the TL structure.

If more than one CS achieve the same similarity score, and they lead to different structural transformations, then the module returns both results as equivalent TL structures. Moreover, if the highest similarity score is lower than a threshold, the input sentence structure is maintained, to prevent transformation towards an ill-fitting prototype. For most of our experiments, an indicative threshold value is between 85 and 90%.

---

[1]For the experiments reported in this volume, the weight values are $W_{phraseType} = 0.6$; $W_{headPoS} = 0.1$; $W_{fheadPoS} = 0.1$; $W_{headCase} = 0.2$.

The algorithm has been extended to tackle the Null-Subject parameter in languages like Greek, using the same alignment information. In the parallel corpus when a subject is not phonologically instantiated from either side of a given corpus sentence, then the phrase containing the subject on the other side will be mapped to an "empty phrase". This allows the algorithm to exploit this information during translation in order to add or remove the subject phrase accordingly in TL.

For a better understanding of this approach, an example is provided next with Greek as the source language and English as the target.

3.2.2 Example of How Structure Selection Works

Let us suppose that we want to translate the Greek sentence (s1) "*Με τον όρο Μηχανική Μετάφραση αναφερόμαστε σε μια αυτοματοποιημένη διαδικασία.*" to English. An exact translation of the input sentence would be "*With the term Machine Translation refer (1st pl) to an automated procedure.*", while the correct English translation is "*The term Machine Translation denotes an automated procedure*".

The input sentence is tagged, lemmatised and chunked using the parsing scheme produced by PAM. The relevant information to the Structure Selection module for the input sentence can be summarised in the structural representation shown in Fig. 3.4.

The input sentence contains four distinct phrases, with the first being a prepositional phrase (phrase type: PP), that has a functional head (fhead) with an "as" tag and a head with a "no_ac" tag, etc.

The next step is to compare the sentence in terms of structure to the sentences from the SL side of the parallel corpus. Let us suppose that one of the sentence pairs from the aligned bilingual corpus is the following (hereafter referred to as (s2)):

Greek: Οι ιστορικές ρίζες της Ευρωπαϊκής Ένωσης ανάγονται στο Δεύτερο Παγκόσμιο Πόλεμο.

English: The historical roots of the European Union lie in the Second World War.

The corpus source sentence has structural information which is described as the following sequence of phrases:

pp(no_nm) pp(no_ge) vg(vb) pp(no_ac)

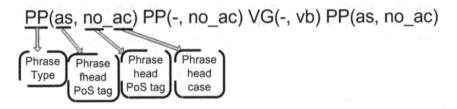

Fig. 3.4 Detailed representation of sentence (s1) with phrase and PoS annotation

Table 3.1 Example of a dynamic programming table when comparing sentences (s1) and (s2)

		Input sentence			
		pp(as, np_ac)	pp(-, no_ac)	vg(-, vb)	pp(-, no_ac)
	0	0	0	0	0
pp(-, no_nm)	0	60	80	-20	60
pp(-, no_ge)	0	60	140	40	40
vg(vb)	0	-50	10	240	140
pp(as, no_ac)	0	100	30	-40	**340**

After calculating the similarity scores for each phrase pair of the input sentence and the SL sentence from the parallel corpus (CS), the dynamic programming algorithm completes a table such as the one depicted in Table 3.1 (with the arrows denoting the highest-scoring aligned sub-sequence).

When an arrow moves diagonally from cell A to cell B, this denotes that the phrases mapped at cell A are aligned. When an arrow moves horizontally, the IS phrase is aligned with a space, and when an arrow moves vertically the CS phrase is aligned with a space.

Table 3.1 forms then the base for calculating the transformation cost (340, i.e. the sum of all transitions), which in this case denotes an 85% similarity score between the two sentences. If during the Structure Selection process, no other SL sentence from the bilingual corpus achieves a higher similarity score, then the aforementioned sentence becomes the basis on which the input sentence will be modified in accordance with the attached TL structure. Using the alignment information of the Dynamic Programming algorithm and the alignment information of the parallel corpus (produced by the Phrase Aligner Module), the phrasal segments of the sentence are reordered accordingly and the sentence is passed on to the second phase of the translation pipeline. It should be noted that even though the Structure Selection module works at the source language level, each word is augmented with the equivalent TL lemmata provided by the bilingual dictionary for subsequent processing.

3.3 Phase Two: Translation Equivalent Selection

The second phase of the translation process attempts to identify the best translation between lexically equiprobable solutions (in terms of the lexicon content). Hence, the Translation Equivalent Selection module performs word translation disambiguation, local word reordering within each syntactic phrase as well as addition and/or deletion of auxiliary verbs, articles and prepositions. In the default settings of PRESEMT, all of the above are performed by only using a syntactic phrase model extracted from the large TL corpus.

The module input is the output of the Structure Selection module. Each sentence contained within the text to be translated is processed separately, so there is no exploitation of inter-sentential information.

The first task is to select the correct TL translation of each word. In the PRESEMT methodology, alternative methods for word translation disambiguation have also been integrated and can be used instead of the default one. These include Self-Organising Maps, n-gram vector space models or SRI n-gram models extrapolated from the TL monolingual corpus, though none of these is used in the PRESEMT configuration reported here.

The second task involves establishing the correct word order within each phrase. With the default settings of the PRESEMT system, this step is performed simultaneously with the translation disambiguation step, using the same TL phrase model. In the case of selecting one of the alternative methods for disambiguation, the phrase model is used only for local word reordering, within the boundaries of the phrases. During word reordering, the algorithm also resolves issues regarding the insertion or deletion of words such as articles and other auxiliary tokens.

Finally, token generation is applied to the lemmas of the translated sentence together with their morphological features. In that way, the final tokens are generated. The token generator used constitutes a simple mapping from lemmas and morphological features to tokens. This mapping has been extracted from morphological and lemma information contained in the monolingual corpus. Due to data

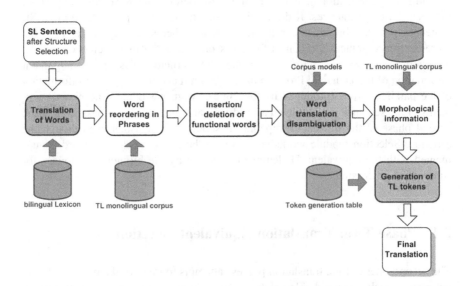

Fig. 3.5 Data flow in the Translation Equivalent Selection module

sparseness, such a mapping will always contain gaps particularly in the case of rather infrequent words. A more sophisticated approach would, therefore, try to close the gaps in the inflectional paradigms of the lemmas. This could for instance be done by inferring inflectional paradigms of infrequent words from those of more frequent words.

Figure 3.5 provides an overview of the Translation Equivalent Selection module, which receives as input the output of the first phase of the main translation engine, i.e. a tagged, lemmatised and parsed source sentence with its constituent phrases reordered in accordance with the target language. The output is the final translation generated by the system:

The two instances of the TL corpus depicted in Fig. 3.5, respectively, refer to the main language model, presented in Chap. 2, and to a table of lemma-token pairs, which is also extracted from the TL corpus.

3.3.1   Applying the Language Model to the Task

When initiating the Translation Equivalent Selection module, a matching algorithm accesses the TL language model to retrieve similar phrases and select the most similar one through a comparison process. The aim is to perform word sense disambiguation and establish the correct word order within each phrase. The comparison process is viewed as an assignment problem that can be solved by using either exact algorithms guaranteeing the identification of the optimal solution or algorithms which yield sub-optimal solutions. In the current implementation, we have opted for the Gale-Shapley algorithm (Gale and Shapley 1962; Mairson 1992), a non-exact algorithm, over the Kuhn-Munkres algorithm (Kuhn 1955; Munkres 1957) that computes an exact solution of the assignment problem, but at a substantially higher cost in terms of computing resources.

The Gale-Shapley algorithm solves the assignment problem by separating the items into two distinct sets with different roles. In this approach, the two sets are termed as (i) suitors and (ii) reviewers. In the present MT application, the aim is to create assignments between tokens of the SL side (which are assigned the role of suitors) and tokens of the TL side (which undertake the roles of reviewers). In the Gale-Shapley algorithm, the suitors have the responsibility of defining their order of preference of being assigned to a specific reviewer, giving an ordered list of their preferences. Based on these lists, the reviewers can each select one of the suitors by evaluating them based on their ordered lists of preference, in subsequent steps revising their selection so that the resulting assignment is optimised. As a consequence, this process provides a solution which is suitor-optimal but potentially non-optimal from the reviewers' viewpoint. However, the complexity of the algorithm is substantially lower to that of Kuhn-Munkres and thus it has been chosen for the Translation Equivalent Selection process as the algorithm of choice, so as to reduce the computation time required. Any errors due to using this sub-optimal approach are limited to the reordering of phrases on the TL side, with no lexical selection changes (since these are decided upon by sampling the indexed files of phrases discussed in Sect. 2.4).

The main issue at this stage is to be able to reorder appropriately any tokens within each phrase, while at the same time selecting the most appropriate translation for each word that the bilingual lexicon has provided. This entails that tokens from (a) a given phrase of the input sentence, call it ISP (input sentence phrase), and from (b) a TL phrase extracted from the TL phrase model and denoted as MCP (monolingual corpus phrase), are close to each other in terms of the number of tokens and type. More specifically, the number and identity of items in a given MCP being used as a template is at least equal to (or larger than) the number of elements in ISP (since it is required to be in a position to handle all tokens of ISP, it is preferable to delete existing MCP elements from their existing locations rather than introducing new ones). In principle, the number of ISP tokens should be equal to or very close to that of MCP. This necessitates a search, which is algorithmically described by the following steps:

Step 1: For each ISP, a decoder creates a vector containing all translation equivalents using the translation alternatives provided by the bilingual lexicon. The number of vectors created is equal to the number of translation equivalents of the phrase head words. The word order does not change.

Step 2: Iteratively process each vector; retrieve for each one the corresponding set of phrases of MCP from the phrase model, based on the phrase type, the lemma and PoS tag of the phrase head.

Step 3: For each retrieved MCP in the vector, apply the Gale-Shapley algorithm for aligning its tokens with the tokens of the ISP. The word alignment provides a guideline for reordering the ISP according to the MCP word order and also provides a similarity score through a comparison formula applied to each one of the aligned word pairs (see the equation below). The similarity score is calculated as the weighted sum of the comparison scores of four types of information, namely (a) phrase type (**PTypeCmp**), (b) phrase head lemma (**LemCmp**), (c) phrase head PoS tag (**TgCmp**) and (d) phrase case (**CsCmp**), if this latter information is available.

$$\text{Score} = w_{\text{ptype}} \times \text{PTypeCmp} + w_{\text{lem}} \times \text{LemCmp} + w_{\text{tag}} \times \text{TgCmp} + w_{\text{case}} \times \text{CsCmp}$$

$$(2.2)$$

where all weights are real positive-valued parameters that sum up to one.

Step 4: After performing all comparisons, the best-matching ISP-MCP pair is selected taking into account the similarity score as well as the MCP frequency of occurrence in the TL corpus. Similarity scores are not compared as absolute values but as a ratio, so as to allow the insertion and/or deletion of words such as articles and other functional words. If the similarity scores of two or more MCPs are close according to the ratio score, then we compare their frequencies in order to determine the best one, and only if the frequencies are also close, we use the absolute comparison values to select the most appropriate ISP-MCP. In this way, a slightly lower scoring solution, which is, however, much more frequent, will be selected over the rarer one.

Step 5: By selecting the most appropriate ISP-MCP pair, the algorithm performs lexical disambiguation by rejecting all other equivalent ISPs in the vector. Words are also reordered based on the MCP using the word alignment produced by the Gale-Shapley algorithm. After applying the previous steps for all phrases in the input sentence, the final sentence translation is produced. It should be noted that the order of phrases has already been established in the Structure Selection module.

Step 6: A token generator component is applied to the lemmas of the TL sentence together with their morphological features. In that way, the final tokens are generated and the final translation is produced.

3.3.2 Example of How TES Works

To illustrate the operation of the Translation Equivalent Selection phase, the handling of the final phrase from the example sentence of Sect. 3.2.2 (*"σε μία αυτοματοποιημένη διαδικασία"*) is presented. This specific phrase comprises four tokens as shown in Table 3.2, where for reasons of simplicity abbreviated SL tags are used, including PoS tag and case only.

For this phrase, the fourth token ("διαδικασία") is the phrase head. Thus, when searching for the best phrase translation, the indexed files for each of the two candidate translations of the phrase head ("procedure" and "process") are searched using as an additional constraint the phrase type (in this case "PC") and the head PoS tag (here "NN"). Hence, following the annotation introduced in Sect. 2.4, files "PC/procedure_NN" and "PC/process_NN", which contain 21,939 and 35,402 distinct entries, respectively, are searched for matching phrase occurrences. Since all four tokens have multiple translations suggested by the lexicon, a number of possible combinations ($6 \times 4 \times 2 \times 2 = 96$) of lemma sequences need to be

Table 3.2 Phrase tokens and their tags in SL and TL, respectively

Token id.	SL token (lemma)	SL PoS tag	Candidate lemmas from bilingual lexicon	TL PoS tag
1	σε (σε)	AsPpSp	[at, in, into, on, to, upon]	IN
2	μια (ένας)	At_Ac	[1, a, an, one]	CD
3	αυτοματοποιημένη (αυτοματοποιημένος)	VbPv_Ac	[automate, automated]	VVN
4 {head}	διαδικασία (διαδικασία)	No_Ac	[procedure, process]	NN

Table 3.3 Candidates of phrase translation retrieved from the TL model

Candidates	Sequence of tokens (lemmatised)	Originating indexed file	TL corpus frequency	Matching score (%)
1	To a store procedure	PC/procedure_NN	3	92.5
2	To an automate procedure	PC/procedure_NN	2	92.5
3	In an ongoing process	PC/process_NN	12	92.5

matched to the phrase instances contained in the two indexed files. Following this matching process, the best-matching phrase instances retrieved from the indexed files are shown in Table 3.3 in order of retrieval.

As can be seen, the first two entries are retrieved from the file containing PC-type phrases with "procedure" as their head, while the third one from the file containing PCs headed with the word "process". An exhaustive search of the two indexed files has shown that no exact matches to the input phrase exist. The highest matching score is 92.5%, as for none of the three examined phrases the lemma of the third token is matched. Still, the 92.5% score is sufficiently high to form a sound basis for the translation (on the contrary if it was below a user-defined threshold typically chosen from the range of 75 to 90%, this translation would be rejected and the SL order of tokens in the phrase would be adopted). In addition, the frequencies of candidates 2 and 3 are comparable, differing by less than one order of magnitude. As all retrieved phrases have equal matching scores, the winning phrase is selected to be the one with the highest frequency of occurrence in the TL monolingual corpus. In the current example, based on the contents of the fourth column, the chosen phrase is the third phrase of Table 3.3. This phrase is then used as the basis for translating the respective SL side phrase, by replacing the token "ongoing" (which is not an appropriate translation, based on the bilingual lexicon) with the token "automated" that is suggested by the lexicon. The sequence obtained with this replacement (namely *"in an automated process"*) represents the translation of this phrase, which forms part of the final sentence translation.

References

Brown P, Cocke J, Della Pietra S, Della Pietra V, Jelinek F, Mercer RL, Roossin P (1988). A statistical approach to language translation. In: Proceedings of COLING'88 Conference, vol 1, pp 71–76

Collins M, Koehn P, Kucerova I (2005) Clause re-structuring for statistical machine translation. In: Proceedings of the Annual meeting of the Association for Computational Linguistics, vol 43, p 531

Gale D, Shapley LS (1962) College admissions and the stability of marriage. Am Math Mon 69:9–14

Hutchins J (2005) Example-based machine translation: a review and commentary. Mach Transl 19:197–211

Koehn P (2010) Statistical machine translation. Cambridge University Press, Cambridge. ISBN: 978-0-521-87415-1

Kuhn HW (1955) The Hungarian method for the assignment problem. Naval Res Logistics Q 2:83–97

Mairson H (1992) The stable marriage problem. Brandeis Rev 12:1. Available at: http://www.cs.columbia.edu/~evs/intro/stable/writeup.html

Munkres J (1957) Algorithms for the assignment and transportation problems. J Soc Ind Appl Math 5:32–38

Nagao M (1984) A framework of a mechanical translation between Japanese and English by analogy principle. In: Elithorn A, Banerji R (eds) Artificial and human intelligence: edited review papers presented at the international NATO Symposium, October 1981, Lyons, France, Amsterdam: North Holland, pp 173–180

Popovic M, Ney H (2006) POS-based word reorderings for statistical machine translation. In: Proceedings of the 5th International Conference on Language Resources and Evaluation (LREC2006), Genoa, Italy, pp 1278–1283

Rottmann K, Vogel S (2007) Word reordering in statistical machine translation with a POS-based distortion model. In: Proceedings of the 11th International Conference on Theoretical and Methodological Issues in Machine Translation (TMI 2007), Skövde, Sweden, pp 171–180

Smith TF, Waterman MS (1981) Identification of common molecular subsequences. J Mol Biol 147:195–197

Chapter 4
Assessing PRESEMT

The topic of the current chapter is the evaluation of the performance of PRESEMT both per se as well as in comparison with other MT systems, the performance relating to the translation quality being achieved. While it is possible to employ humans for this task (subjective evaluation), who assess an MT system in terms of fluency (i.e. grammaticality) and adequacy (i.e. fidelity to the original text) (van Slype 1979), this being a laborious and time-consuming process, evaluation normally relies on automatic metrics (objective evaluation) that calculate the similarity between what an MT system produces (system output) and what it should have produced (reference translation). For PRESEMT, both evaluation processes have been utilised.[1] However, the current chapter is confined to the findings of the objective evaluation, which by being automatically performed can provide the most up-to-date image to the system effectiveness via repetitive applications.

For expository purposes, data from Greek to English will be employed, since evaluation results on the specific language pair have been more extensively studied in the case of PRESEMT. Besides, the fact that there exist key differences between the two languages (Greek being a null-subject language (Biberauer et al. 2010) with free word order in contrast to English, a fixed word order language in which the subject obligatorily is phonologically instantiated) makes the translation task more challenging in comparison with other language pairs.

The chapter is structured as follows: first the dataset is profiled, on which the evaluation was performed. Next, a concise presentation of the automatic metrics used is provided, followed by a section on evaluation objectives and results. Finally, the results acquired for PRESEMT and other MT systems are described and commented upon.

---

[1]Human evaluation has only been carried out once within the PRESEMT project. For the corresponding results, the interested reader is referred to the project deliverable D9.2 (http://www.presemt.eu/files/Dels/PRESEMT_D9.2_supplement.pdf).

© The Author(s) 2017
G. Tambouratzis et al., *Machine Translation with Minimal Reliance on Parallel Resources*, SpringerBriefs in Statistics,
DOI 10.1007/978-3-319-63107-3_4

Table 4.1 Description of the
evaluation dataset

Source language	Greek
Target language	English
Sentence size (in tokens)	7–40
Number of tokens	2758
Number of sentences	200
Domain	None
Number of reference translations	1

4.1 Evaluation Dataset

The evaluation was performed on material collected over the Web according to
certain specifications. More specifically, the Web was crawled over for retrieving a
non-domain-specific Greek corpus of 1000 sentences, the length of which ranged
between 7 and 40 tokens. A subset of 200 sentences randomly selected out of this
corpus constituted the evaluation dataset.

At this point, it should be noted that the specific dataset was compiled prior to
the start of the implementation, to avoid any bias on the actual PRESEMT
experimentation. Moreover, it was intended to be used for development and for a
longitudinal study of the PRESEMT performance; therefore, its size was purposely
kept relatively small.

To obtain reference translations, the whole 200-sentence dataset was manually
translated into English by native speakers of Greek. The correctness of the translations was next verified by native speakers of English, independent from the ones
that originally created the data. The evaluation dataset features are summarised in
Table 4.1.

4.2 Objective Evaluation Metrics

As mentioned before, objective evaluation is prevalent in the MT area since it yields
immediate results and provides a more accurate picture of a system's performance.
It relies on automatic metrics, which measure the similarity of the MT system
output to (usually) one human-produced reference translation. For evaluating
PRESEMT, four metrics have been employed, which are widely used in the field,
namely BLEU, NIST, Meteor and TER.

The BLEU (Bilingual Evaluation Understudy) metric was developed by IBM
(Papineni et al. 2002). Despite the fact that it was intended for evaluating statistical
MT systems, its use is currently widespread in the whole MT field. In essence,
BLEU calculates the number of common n-grams between a system translation and
the reference translation(s). The BLEU score range is between [0, 1], with 1
denoting a perfect match, i.e. a perfect translation.

NIST (2002), functioning similarly to BLEU, also counts the matching n-grams between system translations and reference ones. However, it additionally introduces information weights for less frequent, and for this reason more informative, n-grams. The NIST score ranges between $[0, \infty)$, where a higher score signifies a better translation quality.

Meteor (Metric for Evaluation of Translation with Explicit ORdering), developed at CMU (Denkowski and Lavie 2011), was intended to address weaknesses in BLEU such as the lack of recall (Banerjee and Lavie 2005), hoping to achieve a higher correlation with human judgements. The specific metric calculates the similarity between the system output and each of the reference translations, yielding the highest score of this comparison as its final score. Similarly to BLEU, the score range is $[0, 1]$, where 1 signifies a perfect translation.

TER (Translation Error Rate), developed at the University of Maryland, reflects the philosophy of the Levenshtein distance (Levenshtein 1966), since it calculates the minimum number of edits that are essential for changing the system translation into the reference translation (Snover et al. 2006). If there exist more than one reference translations, then TER takes into account only the reference translation closest to the system translation, since this entails the minimum number of edits. The calculated score, with a range of $[0, \infty)$, derives from the total number of edits, namely insertion, deletion and substitution of single words as well as shifts of word sequences. Hence, a zero score corresponding to zero number of edits denotes a perfect translation.

For the results reported in the current chapter, the following metric versions (with the standard configuration) have been used:

- BLEU and NIST: mteval-v13a;
- Meteor v1.3;
- TER v0.7.25.

4.3 System Evaluation

4.3.1 Evaluation Objectives

The evaluation had a twofold aim: first, the performance of PRESEMT should be assessed not only per se; rather it should be compared to that of other MT systems, in terms of consistency and translation quality. To meet this objective, the same dataset was also translated by two freely available and well-known systems, Google Translate and Bing Translator. The selection of the specific systems, besides their popularity and wide use, was also governed by the fact that they both presumably belong to the SMT paradigm that puts forward the employment of huge bilingual corpora in contrast to the PRESEMT methodology that keeps bilingual resources to a minimum. Thus, it would be interesting to contrast the PRESEMT system to MT systems so different in nature, in terms of their translation performance.

The second aim was to continuously obtain evaluation results, while setting different time points as milestones for all three systems, something which would allow to observe their evolution over time and to form a clear view of PRESEMT's potential improvement in particular. The results presented here are drawn from three different evaluation periods, (a) mid-2012, when the first system prototype of PRESEMT was released (Sofianopoulos et al. 2012), (b) late 2014 (Tambouratzis et al. 2016) and (c) mid-2015.

4.3.2   Evaluation Results

The evaluation scores obtained for the three systems, PRESEMT, Google Translate (indicated as "Google") and Bing Translator (indicated as "Bing"), are given in Table 4.2. For each system, three different versions are identified, corresponding to evaluation milestones, and signified by the equivalent period (e.g. google2012, google2014 and google2015). The improvement or deterioration of performance noted between the first and the second periods and between the second and the third periods is also marked in the table, next to each score.

The "presemt2012" version, which is the first system prototype, expectedly achieves low scores, since it represents a very early release of the system. As one moves to the second evaluation period, a remarkable improvement is noticeable, for the BLEU metric in particular, while the system's performance further advances in

Table 4.2 Evaluation results for PRESEMT, Google and Bing over different time periods

Period	Metric	PRESEMT		Google		Bing	
2012	BLEU	presemt2012		google2012		bing2012	
		0.176		0.554		0.460	
	NIST	5.790		8.805		7.941	
	Meteor	0.336		0.467		0.428	
	TER	61.862		29.791		37.631	
2014	BLEU	presemt2014		google2014		bing2014	
		0.306	73.86%	0.526	−5.05%	0.497	8.04%
	NIST	6.692	15.58%	8.538	−3.03%	8.279	4.26%
	Meteor	0.378	12.50%	0.461	−1.28%	0.452	5.61%
	TER	54.564	11.80%	34.599	−16.14%	34.181	9.17%
2015	BLEU	presemt2015		google2015		bing2015	
		0.366	19.61%	0.522	−0.76%	0.505	1.61%
	NIST	7.119	6.38%	8.575	0.43%	8.295	0.19%
	Meteor	0.405	7.14%	0.465	0.87%	0.453	0.22%
	TER	49.303	9.64%	31.847	7.95%	33.624	1.63%

the third period. This improvement is the outcome of major modifications in certain pre-processing modules (mainly the phrase alignment module) as well as the enhancement of the bilingual lexicon coverage and the enrichment of the token generation module. It is also attributed, especially as regards the "presemt2015" version, to an alternative implementation of the main translation engine algorithm and a more efficient handling of syntactic phenomena.

In comparison with the other two systems, the "presemt2012" version is greatly outperformed by Google and Bing, the translation quality of which is higher by 215 and 161%, respectively (for the BLEU metric). However, as PRESEMT further develops, the differences in translation quality are clearly reduced to 72 and 62% for the second period and to 43 and 38% for the third period. Therefore, the performance of PRESEMT becomes comparable to that of Google and Bing. Although Bing and Google still outperform PRESEMT, achieving high scores (which reflect the substantial resources used to create these systems), there are certain grammatical and syntactic contexts where the two systems are unpredictable in terms of the translation output. Such cases are discussed in the next section.

This lack of consistency is also evident in the results obtained for Google in the second period (and the third period, for the BLEU metric), which denote a decline of the system's performance over time, while the opposite would be expected. A possible explanation for this behaviour could be that Google focuses predominantly on the more widely-used language pairs, and thus the performance of the lesser-used ones (such as Greek to English) may suffer, to a degree.

4.3.3 Expanding the Comparison

After the continuous monitoring of its performance, PRESEMT was also compared for the same dataset to two professional MT systems, SYSTRAN and WorldLingo, which are available online.[2] Interestingly, PRESEMT succeeds in outperforming both systems, SYSTRAN by 24% and WorldLingo by 28% for the BLEU metric. The complete results are given in Table 4.3.

4.3.4 Experimenting with Further Data

The primary positive outcome of the evaluation was that PRESEMT succeeds in achieving a consistent performance. In an attempt to verify that fact, a second 200-sentence evaluation dataset was compiled. This time, however, the 200

---

[2]For our experiments, we have used the online version of SYSTRAN (www.systranet.com/translate) and WorldLingo (www.worldlingo.com/en/products_services/worldlingo_translator.html).

Table 4.3 Comparative evaluation results for PRESEMT, SYSTRAN and WorldLingo

Period	Metric	PRESEMT	SYSTRAN	WorldLingo
2015	BLEU	presemt2015	systran2015	wl2015
		0.366	0.295	0.287
	NIST	7.119	6.486	6.376
	Meteor	0.405	0.383	0.372
	TER	49.303	49.547	50.627

Table 4.4 Description of the second evaluation dataset

Source language	Greek
Target language	English
Sentence size (in tokens)	4–47
Number of tokens	4117
Number of sentences	200
Domain	News
Number of reference translations	1

sentences were longer and were randomly selected out of a corpus of news texts (see Table 4.4). The dataset was translated by the "presemt2015" version, the corresponding versions of Google and Bing as well as SYSTRAN and WorldLingo.

For creating the reference translations, the same process was followed as in the first dataset, namely the dataset was initially translated into English by native speakers of Greek. Subsequently, the translated sentences were validated as to their correctness by native speakers of English.

Table 4.5 lists the evaluation scores obtained for all five MT systems. The consistent behaviour of PRESEMT is verified since it is observed that it maintains the achieved translation quality or slightly exceeds it (in the case of NIST and Meteor metrics). Its performance continues being comparable to that of Google (outperformed by 42% in BLEU), but not to that of Bing, the translation quality of which improves considerably (PRESEMT is outperformed by Bing by 63%).

Table 4.5 Evaluation results for the second evaluation dataset

Period	Metric	PRESEMT	Google	Bing	SYSTRAN	WorldLingo
2015	BLEU	presemt2015	google2015	bing2015	systran2015	wl2015
		0.360	0.509	0.588	0.281	0.265
	NIST	7.409	8.813	9.356	6.505	6.322
	Meteor	0.415	0.461	0.486	0.375	0.358
	TER	57.738	32.080	27.312	50.695	52.132

In comparison with the results of Table 4.3, PRESEMT achieves a still higher translation quality (28 and 36%, respectively, for the BLEU metric) than that of SYSTRAN and WorldLingo, increasing the gap over the earlier dataset.

4.4   Comparing PRESEMT to Other MT Systems

The current section presents specific cases out of the translations produced by the three MT systems, PRESEMT, Google and Bing, in an attempt to sketch a rough picture of their behaviour and also to highlight the fact that PRESEMT, although being outperformed by those two systems in all the evaluation periods, exhibits a consistent performance. This is not to entail that Bing or Google yields a poor output; rather the aim is to lay emphasis on the importance of syntactic/grammatical information for handling certain phenomena in a successful manner.

In the examples that follow, each source sentence is provided together with its correct English translation and the translations produced by the three systems. Errors are denoted in boldface.

The first two examples illustrate that some source words (cf. the possessive clitics in our case) may "appear" or "disappear" in the translation of Bing or Google. PRESEMT, on the other hand, may initially misplace the clitics, but does not fail to include them in the translation.

Example 1

SL sentence: Ο πατέρας της προσπαθεί μάταια να τη μεταπείσει
Translation: Her father tries in vain to dissuade *her*

	2012	2014	2015
PRESEMT	Her father tries vain to her coax	Her father in vain tries to her coax	Her father tries in vain to dissuade her
Google	Her father tries in vain to convince **the** Ø	Her father tries in vain to persuade her	Her father vainly tries to convince Ø
Bing	Her father tries in vain to dissuade Ø	Her father tries in vain to dissuade Ø	Her father tries in vain to dissuade Ø

The Ø symbol indicates missing words

Example 2

SL sentence: Ο δρόμος μας για τη Θράκη και τη Μικρά Ασία περνά από τη Ρωσία
Translation: Our route to Thrace and Asia Minor goes through Russia

	2012	2014	2015
PRESEMT	Our way for Thrace all little Asia spends of Russia	Our way for the minor Thrace and Asia goes of Russia	Our way about Thrace and the minor Asia goes of Russia

(continued)

(continued)

SL sentence: Ο δρόμος μας για τη Θράκη και τη Μικρά Ασία περνά από τη Ρωσία
Translation: Our route to Thrace and Asia Minor goes through Russia

	2012	2014	2015
Google	The way **our** Thrace and Asia Minor passes from Russia	**Our** route to Thrace and Asia Minor passes from Russia	Ø The road to Thrace and Asia Minor passes through Russia
Bing	Ø The road to Thrace and Anatolia passed by Russia	**Our** road to Thrace and Asia minor passes from Russia	Ø The road to Thrace and Asia minor passes from Russia

Greek is a null-subject (NS) language. This means that the subject of the finite verb is not obligatorily expressed. In English, on the other hand, which is a non-null-subject (NNS) language, the subject of a finite verb must always be phonologically instantiated. A syntactic structure-aware system such as PRESEMT may succeed in capturing this major distinction between the two languages as opposed to Google or Bing. In the following examples, the pronominal subject is either not generated (Example 3) or does not appear in all periods (Example 4) in the case of Google and Bing.

Example 3

SL sentence: Πιστοποιούν την ελληνική ταυτότητα για το αύριο
Translation: They verify the Greek identity for tomorrow

	2012	2014	2015
PRESEMT	Ø Verified Greek identity for tomorrow	Ø Verify the Greek identity for the tomorrow	They verify the Greek identity for tomorrow
Google	Ø Certify the Greek identity for tomorrow	Ø Certify the Greek identity for tomorrow	Ø Certify the Greek identity for tomorrow
Bing	Ø Showing the Greek identity for tomorrow	Ø Certify the Greek identity for tomorrow	Ø Certify the Greek identity for tomorrow

Example 4

SL sentence: Και αρχίζουν να λένε πόσο περίπλοκα είναι τα πράγματα
Translation: And they begin to say how complex things are

	2012	2014	2015
PRESEMT	All to start Ø call how complex are things	And to tell Ø begin how complex are the things	And they begin to tell how complex are the things
Google	And they begin to say how complex it is	And Ø begin to say how complicated things are	And they begin to say how complex it is
Bing	And Ø begin to say how things are complicated	And Ø begin to say how complicated things are	And Ø begin to say how complicated things are

In a similar vein, subject-verb agreement requires being aware of the syntactic structure. The following examples indicate that in most cases PRESEMT is successful in handling such phenomena, while Google and Bing make errors either in the pronominal subject (Example 5) or in the verb form (Example 6).

Example 5

SL sentence: Αν διαβάσουμε ιστορία θα καταλάβουμε γιατί δεν έχει γίνει Εθνικό Κτηματολόγιο
Translation: If we read history we will understand why there has been no National Cadastre

	2012	2014	2015
PRESEMT	If the book ∅ read history because ∅ not will understand has become the ethnic register land	If we read history we will comprehend because has not been national land register	If we read history we will understand because has not been national land register
Google	If we read history ∅ will understand why there has been no National	If **you** read history **you** will understand why **he** has become National Cadastre	If **you** read history we will understand why **it** has not become the National Cadastre
Bing	If **you** read history **you** will understand why there has been no National Register	If **you** read history **you** will understand why **he** has not become a national cadastre	If **you** read history **you** will understand why **he** has not become a national cadastre

Example 6

SL sentence: Ο λαός, σ' αυτό το νοητικό και πραγματικό σχήμα, παίζει καταλυτικό ρόλο
Translation: The people, in this mental and actual shape, play a catalytic role

	2012	2014	2015
PRESEMT	The people, in the true and intellectual figure, plays a decisive role	The people, in this intellectual and actual format, plays catalytic part	The people, to this intellectual and real figure, plays catalytic role
Google	The people in this mental and real shape, plays a catalytic role	The people in this mental and actual shape, **playing** a catalytic role	The people, in this mental and actual shape, **playing** a catalytic role
Bing	The people in this mental and real shape, plays a catalytic role	The people, in this mental and physical shape, plays a catalytic role	The people in this mental and physical shape, plays a catalytic role

Furthermore, features such as gender or number can also be mistranslated, as the following examples indicate, where Google and Bing fail to produce the correct form of the personal pronoun.

Example 7

SL sentence: Έχω ζήσει μ' αυτήν σ' ένα σκοτεινό, κρύο και υγρό δωμάτιο
Translation: I have lived with her in a dark, cold and damp room

	2012	2014	2015
PRESEMT	have lived in she in mysterious, wet and room cold	I have lived to her in a dark, cold and wet room	I have lived to her in a dark, cold and wet room
Google	I've lived with **it** in a dark, cold and damp room	I've lived with **it** in a dark, cold and wet room	I live with **it** in a dark, cold and damp room
Bing	I have experienced **it** in a dark, cold and wet room	I've lived with **it** in a dark, cold and wet room	I've lived with **it** in a dark, cold and wet room

4.5   Conclusions

In the current chapter, the results of the evaluation of the PRESEMT system were presented and discussed. Automatic metrics were employed, which produce a comparison score calculated on the similarity between a reference translation and a system translation. The evaluation was carried out in three different periods in time, thus allowing the study of the evolution of PRESEMT over time. The translation quality attained by PRESEMT was also compared to that of Google Translate and Bing Translator, two widely-used translation systems, as well as SYSTRAN and WorldLingo, which are professional MT systems.

The first conclusion drawn is that PRESEMT consistently achieves an improved performance over time, which is reflected in the evaluation scores but is also evident in the translated text itself. Secondly, although outperformed by the Google and Bing, it succeeds in conveying the gist of the translation, something specified in the design of the system. At the same time, while being an MT system with a brief development history, it manages to produce better translations than the ones produced by professional systems like SYSTRAN or WorldLingo. Thirdly, in several cases, PRESEMT achieves a comparable quality to that of Google or Bing, although it relies on a significantly smaller bilingual resource. Fourthly, the fact that PRESEMT operates on syntactic entities (phrases) allows it to exhibit a predictable behaviour, in contrast to Google or Bing, in handling certain phenomena (e.g. null-subject structures or subject-verb agreement), thus highlighting the importance of syntactic information. Finally, it is indicated that large bilingual resources are not essential in achieving a sufficient translation quality.

The above leads us to consider that PRESEMT has a promising future, while further improvements are attainable by appropriately modifying specific system modules. This is discussed further in Chaps. 6 and 7.

References

Banerjee S, Lavie A (2005) METEOR: An Automatic Metric for MT Evaluation with Improved Correlation with Human Judgments. In: Proceedings of Workshop on Intrinsic and Extrinsic Evaluation Measures for MT and/or Summarization at the 43rd Annual Meeting of the Association of Computational Linguistics (ACL-2005), Ann Arbor, Michigan, pp 65–72

Biberauer T, Holmberg A, Roberts I, Sheehan M (2010) Parametric variation: null subjects in minimalist theory. Cambridge University Press

Denkowski M, Lavie A (2011) Meteor 1.3: automatic metric for reliable optimization and evaluation of machine translation systems. In: Proceedings of the EMNLP 2011 Workshop on Statistical Machine Translation, Edinburgh, Scotland, pp 85–91

Levenshtein VI (1966) Binary codes capable of correcting deletions, insertions, and reversals. Sov Phys Dokl 10:707–710

NIST (2002) Automatic evaluation of machine translation quality using n-gram co-occurrences statistics (available at: http://www.itl.nist.gov/iad/mig/tests/mt/doc/ngram-study.pdf)

Papineni K, Roukos S, Ward T, Zhu WJ (2002) BLEU: a method for automatic evaluation of machine translation. In: Proceedings of the 40th Annual Meeting of the Association for Computational Linguistics, Philadelphia, U.S.A., pp 311–318

Snover M, Dorr B, Schwartz R, Micciulla L, Makhoul J (2006) A study of translation edit rate with targeted human annotation. In: Proceedings of the 7th AMTA Conference, Cambridge, MA, USA, pp 223–231

Sofianopoulos S, Vassiliou M, Tambouratzis G (2012) Implementing a language-independent MT methodology. In: Proceedings of the 1st Workshop on Multilingual Modeling (held within ACL-2012), Jeju, Republic of Korea, pp 1–10

Tambouratzis G, Vassiliou M, Sofianopoulos S (2016) Language-independent hybrid MT: comparative evaluation of translation quality. Chapter. In: Hybrid Approaches to Machine Translation, Costa-jussà, M.R., Rapp, R., Lambert, P., Eberle, K., Banchs, R.E., Babych, B. (Eds.). Springer-Verlag, pp 131–157. ISBN 978-3-319-21311-8.

van Slype G (1979) Critical study of methods for evaluating the quality of machine translation. Technical Report BR19142, Bureau Marcel van Dijk/European Commission (DG XIII), Brussels (available at: http://issco-www.unige.ch/projects/isle/van-slype.pdf)

Chapter 5
Expanding the System

Following the detailed description of the PRESEMT Machine Translation system and the report on its performance, the current chapter focuses on the system's portability. Portability is a term intended to signify the process of integrating a new language pair into the system. This involves reviewing all the necessary system modules and resources and making all the necessary modifications.

PRESEMT has been designed to be easily extensible and is based on a methodology that does not vary across languages; yet it has proven effective to make certain provisions depending on the language pair at hand. For example, the parameter of agreement should be taken into account when translating into a morphologically rich language.

To exemplify the process, two cases of integration will be discussed involving a reverse language pair (English to Greek) and a totally new one (Greek to German).

Regarding the English to Greek language pair, recall that in Chap. 4 (Evaluation), where results from the reverse language pair were reported, it was claimed that the translation process becomes challenging when moving from a null-subject language (Greek) to a non-null-subject language (English), since, while lacking a phonologically instantiated SL subject, it is obligatory to generate a subject in the TL. Furthermore, the free constituent order in Greek makes the ordering of the corresponding English phrases a more demanding task. Handling the opposite direction additionally entails accommodating agreement in an appropriate way, since Greek is a morphologically rich language.

Greek to German is also of interest since both languages are morphologically rich, while they are characterised by the null-subject versus non-null-subject contrast.

The chapter is structured as follows: the first part recounts the steps needed for adding language pairs into the system. Then, language-pair-specific cases are discussed.

© The Author(s) 2017
G. Tambouratzis et al., *Machine Translation with Minimal Reliance on Parallel Resources*, SpringerBriefs in Statistics, DOI 10.1007/978-3-319-63107-3_5

5.1 Preparing the System for New Language Pairs

The current section recounts the steps followed for PRESEMT to handle (a) a
completely new language pair (Greek to German) and (b) a reverse language pair
(English to Greek). Each of these cases is separately discussed per resource and
system module.

Bilingual lexicon: A new bilingual resource is added to the system, the Greek to
German bilingual lexicon in this case. This lexicon must be processed, so that it is
rendered in the proper format and contains only the necessary information, i.e.
lemma and tag in both the source and the target languages. If the resource is already
part of the system, as was the case with the Greek to English lexicon, then no action
is needed, since the PRESEMT lexica are read bidirectionally.

Bilingual corpus: For compiling the Greek-German corpus, a selection of 200
sentences was made out of a bilingual resource drawn from the Web. In the case of
English to Greek, the existing corpus was reused.

Monolingual corpus: Adding Greek and German to the system as new target
languages entailed compiling the corresponding monolingual corpora. These
resources were automatically created via Web crawling. Please refer to Chap. 2 for
more information regarding the monolingual corpora used in PRESEMT.

Annotation of the corpora: The annotation levels are dependent on whether a
language is a source language or a target one. In the SL case, only tag and lemma
information is needed, whereas the TL side additionally requires phrase segmentation. Table 5.1 details the annotation per corpus and language.

Creating wrappers: System-external tools (e.g. taggers) need to be integrated into
the system via appropriate wrappers so that they can be called during the translation
process. In the cases that are examined in the present chapter, no action was needed,
since the corresponding tools were already integrated.

Table 5.1 Annotation of required corpora per language pair

Language	Type	Tool employed	Corpus	Annotation level
English	SL	TreeTagger (Schmid 1994)	• SL side of the bilingual corpus	Tag and lemma
German	TL	TreeTagger and RFTagger (Schmid and Laws 2008)	• TL side of the bilingual corpus • Monolingual corpus	Tag, lemma, phrase
Greek	SL	FBTagger	• SL side of the bilingual corpus	Tag and lemma
Greek	TL	FBTagger and ILSP parser (Prokopidis et al. 2011)	• TL side of the bilingual corpus • Monolingual corpus	Tag, lemma, phrase

Defining heads: For any language pair integrated into the system, either a new or a reverse one, the heads of phrases must be defined. This is done manually and is governed by the given source language.

Making things parallel: In order to be parallelised, the bilingual corpora were processed by the Phrase Aligner Module (see Chap. 2), which (a) extracts alignments at word and phrase level and (b) segments the SL side of the corpus into phrases. Moreover, PAM automatically calculates SL-TL tag correspondences out of the bilingual lexica.

Creating phrase models: Another step towards integration involves generating a phrase model for the source language that will be used during the translation process for segmenting the SL input into phrases. This is implemented by invoking the relevant module from the PRESEMT package for the PAM/PMG training via a sequence of just two instructions.

Token generation: The generation of tokens (specific forms) out of lemmata involves (a) establishing correspondences of tags between the source and the target languages and (b) drawing up tables of token-tag-lemma triplets to be used during the translation. For these tasks, there are dedicated system modules, which process the bilingual corpora and the monolingual ones for, respectively, extracting (a) SL-TL tag correspondences and (b) token generation tables. These correspondences are extracted in an automated manner and can afterwards be manually enriched.

Word sense disambiguation: This module, already existing in the system, was employed for processing the monolingual corpora in order to extract word frequencies-of-occurrence. These frequencies are used for disambiguation during the second translation phase, as discussed in Chap. 3.

Main translation engine: The core system engine was subjected to no modifications thanks to its language-independent nature.

From the list above, it is deduced that only the system resources need enrichment, while pre-processing modules are reused and the core system engine stays intact. In the next section, the question of handling language-pair-specific features will be discussed.

5.2   Examining Language-Pair-Specific Issues

As was shown before, PRESEMT has a language-independent methodology that accounts for the easy portability of the system. Yet there are cases related to the language pair at hand that require the development of peripheral modules to handle them. These cases are presented in the rest of this section. Such cases are indicative, but those listed herewith provide a sufficient coverage of phenomena for the chosen languages, to supplement the translation phenomena being covered by the PRESEMT algorithms.

5.2.1   Agreement Within a Nominal Phrase

The constituent words of a nominal phrase usually[1] agree in case and number features. When the target language is for instance English, this agreement is not morphologically instantiated.

However, when translating into a morphologically rich language like German or Greek, a module must check agreement within nominal phrases and make the agreement features of the phrase head percolate to the residual words of the phrase such as the article or the modifier (cf. the following examples where the phrase head is marked in bold). Note that the required module is not developed only for one language but for all languages exhibiting agreement.

		Example	Agreement features
SL	English	the faithful **husbands**	Case: Accusative
TL	Greek	τους πιστούς **συζύγους**	Number: Plural
SL	Greek	του πιστού **συζύγου**	Case: Genitive
TL	German	des treuen **Ehemannes**	Number: Singular

5.2.2   Case Mismatches

In the languages that we examine, case differences may emerge usually in the object of the verbs. Likewise, there may be differences in the phrase type of the object (prepositional phrase vs. noun phrase). This phenomenon can be handled by resorting to a verb-object model automatically extracted from the TL monolingual corpus that provides information about the correct case of the object.

		Example	Agreement features
SL	English	The question relates to the **substance** of the matter	Case: Accusative
TL	Greek	Το ερώτημα άπτεται της **ουσίας** του θέματος	Case: Genitive
SL	Greek	Θα συμμετάσχει στην **έκθεση**	Case: Accusative
TL	German	Er wird an der **Messe** teilnehmen	Case: Dative

5.2.3   The Null-Subject Parameter

The parameter at hand broadly distinguishes languages into two categories: (a) non-null-subject languages where the pronominal subject of the finite verb

---

[1]This is not the case with words like, for example, adverbs which lack agreement features altogether (cf. "ο **πάνω** όροφος" = "the floor **above**").

obligatorily is phonologically instantiated (e.g. English and German) and (b) null-subject languages like Greek where the pronominal subject of the finite verb is usually null. The most challenging task is translating from an NS language (Greek in our case) into an NNS one (German), since a module is needed for generating a subject based on the features of the verb with which the subject agrees in person and number.

		Example	Language type	Agreement features	Comment
SL	English	**They** kill 270 and injure more than 750	NNS	Person: Third	The subject is translated and
TL	Greek	**Αυτοί** δολοφονούν 270 και τραυματίζουν περισσότερους από 750	NS	Number: Plural	may optionally remain
SL	Greek	Ø Είχα πιει πολύ το προηγούμενο βράδυ	NS	Person: First	A subject must be obligatorily
TL	German	Die Nacht davor hatte **ich** viel getrunken	NNS	Number: Singular	generated

5.2.4 Word Order

Greek is a free word order language in contrast to English and German, which are characterised by a fixed word order. This type of difference is handled by the Structure Selection module of the main translation engine, which is responsible for correctly ordering constituents.

		Example	Comment
SL	English	Her father **always** wished for a son	Positioning of the adverb before the verb
TL	Greek	• Ο πατέρας της επιθυμούσε **πάντα** ένα γιο • Ο πατέρας της **πάντα** επιθυμούσε ένα γιο	Positioning of the adverb before or after the verb
SL	Greek	Το 1902 απέμεινε μία μόνο καλόγρια που **περιποιούνταν** το ναό	Usually a VO order in the secondary clauses
TL	German	Im Jahr 1902 blieb nur eine Schwester, die sich um den Tempel **kümmerte**	Positioning of the finite verb at the end of a secondary clause

5.3 Notes on Implementation

The present chapter discusses work in progress which has continued after the completion of the PRESEMT project. This work has been carried out within a national project which has formed the continuation effort of the PRESEMT methodology (under the auspices of the Polytropon project). As such, a large portion of this work lies beyond the scope of the current volume. Within this project, a substantial improvement has been recorded for the Greek to German language pair in terms of metrics such as BLEU and NIST, approaching 0.09 and 4.30, respectively, with further scope for improvement.[2]

During the implementation effort, one limiting factor was found to be the quality of the lexicon, as (i) a very large number of non-literal translations were provided and (ii) many multi-word translations were included. Experimentation showed that a cleaning up of the lexicon can substantially improve the performance.

Besides, the fact that German features compound words with very high frequency has an impact on the translation quality (cf. several Greek lemmas correspond to a single word in German, for instance "δυσανεξία στην τροφή" ["food intolerance"] is translated as "Nahrungsmittelunverträglichkeit"). The ability to detect accurately and effectively 1-to-n correspondences has been added in the PRESEMT mechanism via algorithmic modifications. This modification has proved particularly effective, especially in supporting the operation of the Phrase Aligner Module (PAM). Based on this, the improved accuracy of token correspondences allows a more accurate discovery of structural transformation of phrases from source to target language.

5.4 Conclusions

The current chapter described the process of adding a new language pair into the PRESEMT system. Two different cases were examined, a pair of languages completely new to the system and the opposite direction of an already existing language pair. The first case was exemplified with the Greek to German language pair, while the second one involved English to Greek. Both cases shared the interesting features of (a) the target language being a morphologically rich one and (b) the language pair involving the contrast null-subject versus non-null-subject. The integration process was recounted stepwise, considering all the system resources and modules that need to be enriched or modified.

It has been shown that for PRESEMT to handle a new language pair, only additions in terms of resources (lexica, bilingual and monolingual corpora, etc.) are needed, while pre-processing modules (e.g. the Phrase Aligner Module) can be

---

[2]Development and Optimisation of Translation Technology. Deliverable D2.6.2 of Polytropon project, Athens 30/6/2015 (available in Greek at www.presemt.eu/continuation/).

reused and, most importantly, the main translation engine remains the same. That is an additional proof of the language-independent design and implementation of the system. Although specific provisions are rendered necessary for treating various phenomena (e.g. agreement), these can be handled by peripheral modules, thus not altering the core system engine.

The above also supports the view that PRESEMT offers an eminently suitable methodology when developing an MT system for a language pair with limited resources.

References

Prokopidis P, Georgantopoulos B, Papageorgiou H (2011) A suite of NLP tools for Greek. In: Proceedings of the 10th International Conference on Greek Linguistics (ICGL10), Komotini, Greece, pp 373–383

Schmid H (1994) Probabilistic part-of-speech tagging using decision trees. In: Proceedings of the International Conference on New Methods in Language Processing, Manchester, UK, pp 44–49

Schmid H, Laws F (2008) Estimation of conditional probabilities with decision trees and an application to fine-grained POS tagging. In: Proceedings of the 22nd International Conference on Computational Linguistics (COLING 2008), Manchester, UK, pp 777–784

Chapter 6
Extensions to the PRESEMT Methodology

This chapter describes a number of improvements performed on the basic PRESEMT system. These improvements are aimed at specific modules of the system in an effort to achieve gains in the translation accuracy, for which alternative implementations have been suggested. These extensions concern different modules of the PRESEMT architecture. The first extension covers the pre-processing stage, where an improved phrasing model for the SL side is proposed. The second extension involves the use of supplementary language models (LM) in the TL, to improve the translation accuracy in terms of both the phrasal level but also the post-editing and token generation steps.

6.1 Splitting SL Sentences into Phrases More Accurately

Within the PRESEMT approach, one critical constraint is the volume of available training data in terms of parallel corpora. This data is the sole information supporting the transformation from SL to TL in terms of phrase sequences. To enhance portability to new language pairs, this parallel data needs to be very limited. PRESEMT utilises a parallel corpus of a few hundred sentences, supplemented by a large TL monolingual corpus from which a Target Language Model (TLM) is extracted.

To establish phrases, in PRESEMT a TL parser processes the TL sentences (including the TL side of the parallel corpus and the TL corpus). The present section describes efforts to extrapolate the best possible module that will split SL text into phrases (this module is hereafter termed chunker), which needs to be matched to the TL parser. This SL chunker is extracted from the parallel corpus by incorporating TL side parsing information, so that the two phrasing schemes are compatible (i.e. the same types of chunk are established in both SL and TL).

The work reported here relates closely to cross-language approaches transferring phrasing schemes from one language to another, to supplement the sparse data

© The Author(s) 2017
G. Tambouratzis et al., *Machine Translation with Minimal Reliance on Parallel Resources*, SpringerBriefs in Statistics,
DOI 10.1007/978-3-319-63107-3_6

available. Several studies involving the transfer of phrasing schemes across languages have been proposed, extrapolating information from a resource-rich to a resource-poor language. Those include work by Yarowsky and Ngai (2001), Hwa et al. (2005) and Smith et al. (2009).

The default chunker module (PMG) implementation, as reported in Chap. 2, utilises the CRF stochastic model (Lafferty et al. 2001; Wallach 2004). CRF is widely regarded as the model of choice for the generation of parsers (e.g. Sha and Pereira 2003; Tsuruoka et al. 2009), when the environment of a phrase (i.e. neighbouring phrases in the content of a sentence) needs to be modelled. The main drawback of CRF is its substantial number of internal parameters that requires a large volume of training patterns for accurate model extraction. However, employing a large parallel corpus in PRESEMT would compromise its portability to new language pairs.

Though CRF has demonstrated a competent performance, the question is whether a different phrasing model can be trained more effectively on this training data and achieve a better translation. Thus, in later stages of the development of PRESEMT, an alternative methodology was developed for building an SL parser. This uses a mathematically less complex system based on a template-matching approach (denoted as TEM) that has been studied for segmenting text-to-be-translated into linguistically motivated phrases. TEM creates a look-up ordered table of phrases, where for each distinct phrase pattern (determined by the TL parser) the likelihood of correct identification is calculated.

6.1.1 Design and Implementation of TEM

TEM adopts a learn-by-example training, recognising phrases encountered in the training set. This approach is based on the template-matching algorithm described in (Duda et al. 2001), the principle of which is to match part of the input sentence to a known phrase archetype. TEM (i) does not generate a high-order statistical phrasing model taking into account the wider phrasal context and (ii) cannot revise decisions so as to reach a global optimum. Instead, it implements a greedy search (Black 2005) and thus may result in sub-optimal solutions. On the other hand, its learn-by-example process is substantially less complex than the CRF training process, while the model is also easier to fine-tune.

TEM is trained on the SL side sentences of the bilingual corpus, segmented into phrases. Training consists of four steps:

1-Accumulate: From the bilingual corpus of chunked SL sentences, a list of occurring phrases is compiled together with the frequency of occurrence.
2-Order: Phrases are arranged in descending order of likelihood of occurrence, based on an ordering criterion.

3-Expand: The phrase list is expanded, to increase the actual coverage of the phrase space. Each phrase where all declinable words share the same case is generalised to all other cases.

4-Remove: Phrases that are grammatically inconsistent are discarded. Such phrases include those containing inflected words bearing different cases.

The latter two steps integrate a limited amount of language-specific knowledge to enhance the TEM operation. An example of the creation of a phrasing model from a small set of two chunked sentences is shown in Fig. 6.1, showing the

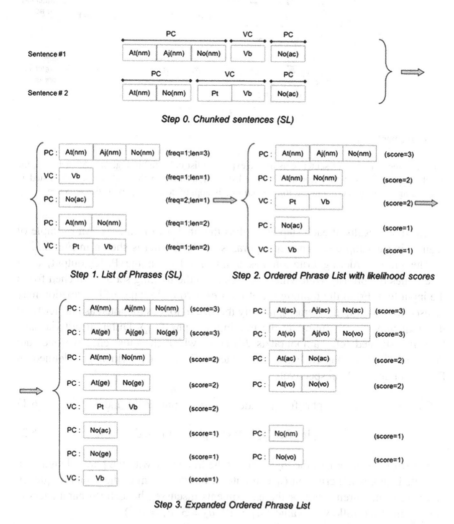

Fig. 6.1 Creation of phrase templates via TEM from a set of two chunked sentences, depicting intermediate results of steps 1, 2, 3 (no ungrammatical phrases exist in this corpus)

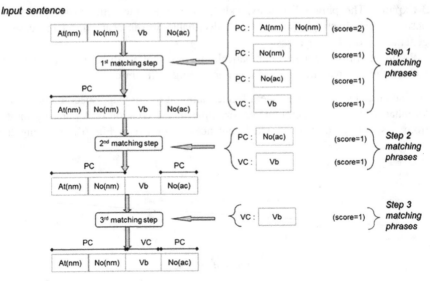

Fig. 6.2 Segmentation of sentence using the expanded list indicated in Fig. 6.1. At each step, the matching phrases from the expanded phrase list are indicated on the *right-hand side*, listed in descending order of score and the top-ranked one is applied to cover part of the sentence

intermediate results at each step as well as the final list of phrases. An example of segmenting a sample sentence into phrases by this model is shown in Fig. 6.2.

The phrase-ordering criterion has a major effect in the PMG output, as it determines the priority with which phrases from the training set are searched for in the input text. From the training set of phrases, two main types of information may be used for ordering the phrases, namely the size in tokens (denoted as phr_len) and the frequency of occurrence in the training set (phr_freq). Different criteria have been investigated (cf. Tambouratzis 2014), of which the most effective ones are discussed in this chapter. These are criteria C_D and C_F, which are defined in Eqs. (6.1) and (6.2), respectively.

$$C_D = \max\{\text{floor}[log(\text{phr\_freq}[\text{p\_index}]) * a_1] + \text{phr\_len}[\text{p\_index}] * a_2\} \quad (6.1)$$

$$C_F = \max\{\text{phr\_freq}[\text{p\_index}] + \text{phr\_len}[\text{p\_index}] * a_4\} \quad (6.2)$$

Criterion C_D combines the logarithm of the frequency with the phrase length in a weighted sum, while criterion C_F computes a weighted sum of the actual frequency and length measurements (for the experiments discussed henceforth parameters a_i are assigned the following values: $a_1 = 10$, $a_2 = 3$, $a_4 = 10^3$).

6.1.2   Experimental Evaluation

The proposed TEM approach is evaluated by studying its effects in the final translation result, to allow the use of objective metrics. For the given language pair (Greek to English), the phrase types are defined by the TL parser (TreeTagger in our case) and thus include ADVC, ADJC, PC and VC. In addition, a fifth phrase type has been defined to indicate tokens which are isolated (i.e. are not included in a phrase), these being defined as ISC (standing for isolated word chunk). Phrase types are considered distinct, and thus the type (in addition to the boundaries) of each phrase must be correctly identified during chunking to achieve a correct segmentation. An example of two sentences that have been parsed into phrases is shown in Table 6.1.

To train the phrasing model, the standard set of 200 parallel sentences (the bilingual corpus of PRESEMT) is used. Indicatively, the number of phrase templates established is depicted in Table 6.2, for the different phrase types. As can be seen, apart from the main phrase types (PC, VC, ADJC and ADVC), TreeTagger also generates three minor ones (CONJC, INTJ and PRT). In addition, for isolated words the ISC phrase type is introduced. In the second row of the table, the number of phrase patterns for each type is depicted. The largest increases are for phrase

Table 6.1 Parsing of two sentences, by (i) the human specialist, (ii) CRF and (iii) TEM with criterion C_F

Model	Segmentation into phrases
Gold (human)	PC1(Ο λαός), PC2(σε αυτό το νοητικό και πραγματικό σχήμα), VC3 (παίζει) PC4(καταλυτικό ρόλο) [PC1(The people), PC2(in this mental and actual shape), VC3(play) PC4(a catalytic role)]
CRF	PC17(Ο λαός), PC18(σε αυτό το νοητικό και πραγματικό σχήμα), VC19 (παίζει) PC20(καταλυτικό ρόλο) [PC17(The people), PC18(in this mental and actual shape), VC19(play) PC20(a catalytic role)]
TEM + C_F	PC2(Ο λαός), PC6(σε) PC8(αυτό) PC10(το νοητικό και πραγματικό σχήμα), VC17(παίζει) PC19(καταλυτικό ρόλο) [PC2(The people), PC6(in) PC8(this) PC10(mental and actual shape), VC17 (play) PC19(a catalytic role)]

Table 6.2 Number of phrase templates established by original and expanded TEM models

Phrase type	PC	VC	ISC	ADJC	ADVC	PRT	CONJC	Total
Original TEM model (step 2)	266	23	17	11	10	2	1	330
Expanded TEM model (step 3)	625	48	25	27	16	2	1	744
Change (%)	134.97	108.7	47.06	145.45	60.0	0	0	125.45

Table 6.3 Translation accuracy for Greek to English, with TEM and CRF, using an improved
PAM output

PMG version	BLEU	NIST	METEOR	TER
CRF	0.335	7.098	0.385	51.027
TEM + C_D	0.340	7.052	0.390	51.096
TEM + C_F	0.347	7.168	0.394	49.983

types which involve cases (namely PC and ADJC). The population of these classes
is more than doubled, as is the case for the total number of phrase patterns in the
TEM model.

Comparative results for CRF and the highest-scoring TEM configurations are
listed in Table 6.3. For CRF, the best configuration established in Tambouratzis
et al. (2011) has been used. Based on the results quoted in Table 6.3, the performance of TEM remains superior to that of the CRF-based system for all four MT
metrics. For instance, CRF leads to a BLEU score of 0.335, while TEM exceeds
that score by more than 3%. Furthermore, for the given experiments, only TEM
achieves a TER score below the 50.0 level. On the whole, TEM provides a wel-
come improvement in quality over the established CRF approach.

The effect of the parallel corpus size on the translation accuracy has also been
studied, by using subsets of the standard 200-sentence parallel corpus. Thus, new
simulations were performed using (i) the first 50 and (ii) the first 100 sentences. The
translation accuracy as expressed by metrics BLEU and TER is depicted in
Fig. 6.3a, b, respectively. As a rule, the performance of PRESEMT using TEM
remains virtually constant, while the corpus size affects more the CRF performance.
For CRF, BLEU increases by approximately 15% when augmenting the parallel
corpus from 50 to 200 bilingual sentences. Similarly, the improvement of TER is
5%. These results indicate that the larger corpus allows a more successful estimation of the CRF model parameters, promising a further improved performance
for larger bilingual corpora.

6.1.3 Conclusions

A template-matching approach has been used as a foundation for creating an
alternative phrasing model for SL text, replacing the standard CRF-based module.
Trained on the same training data as the CRF model, the TEM model has the
advantage of a much more transparent operation. The best TEM configurations are
superior to those of CRF for all MT-specific objective metrics reflecting translation
accuracy. This indicates that the simple TEM algorithm has a sufficiently high
functionality, and at the same time provides a more suitable match to the limited
training set of valid phrases.

A further TEM advantage is its greater ease of enhancement in the pursuit of a
better performance, due to the increased transparency of the phrasing process. The

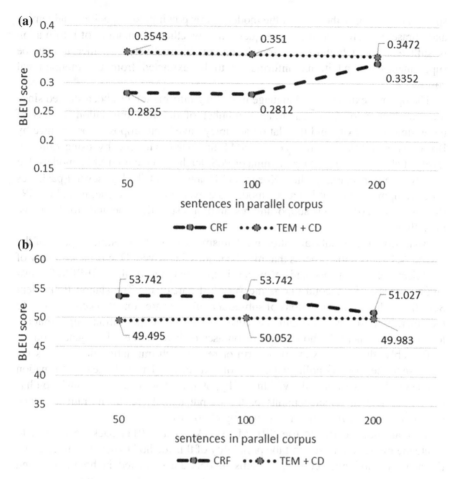

Fig. 6.3 Comparison of translation accuracies, as reflected by (a) BLEU and (b) TER, using for chunking TEM and CRF, for different sizes of the parallel corpus

more easily modifiable TEM nature may provide the foundation for further improving the translation accuracy of PRESEMT in the future.

6.2 Combining Language Models of Different Granularity

As PRESEMT is a phrase-based translation system, one of the key modules is Stage Two (TES) of the translation process, which translates sub-sentential parts of the sentence that correspond to a single phrase, processing one part at a time. In this case, PRESEMT consults a language model (LM) which has been generated by creating phrasal models for each type of phrase (cf. Sect. 2.4). When translating a specific phrase, only the parts of the models corresponding to a specific head lemma and phrase type are consulted. The question is whether this amount of information is sufficient for a high-quality translation, or, alternatively, whether it may be supplemented by additional information to be extracted from the monolingual corpus.

The optimal extraction of language modelling information has been studied since the introduction of statistical MT. A number of different translation models of increasing complexity and translation accuracy have been proposed for instance by Brown et al. (1993). More recently, SMT has been enhanced by using different levels of abstraction, e.g. word, lemma or PoS, leading to factored SMT models that improve the translation quality (Koehn and Hoang 2007). Today, several packages for developing statistical language models are available for free use, including SRI (Stolcke et al. 2011), thus supporting research into statistical methods for language modelling.

Apart from the actual translation mechanism, there are two distinct aspects of the PRESEMT translation process that may be readily improved. One is the accuracy of the token generation mechanism, which is the final step of the PRESEMT translation process. The second one is the very last part of the Translation Equivalent Selection, where the translations of subsequent phrases are concatenated one next to the other. In these cases, the aim is to correct errors in the translation output that are localised either near the boundaries of phrases or involve the token generation.

Probably, the easiest way to confirm or revise such translation decisions is via recourse to the large monolingual corpus of texts in TL. This volume of information can be used to investigate the validity of hypotheses that the correct translation has been generated in selected points of the output text. Here, an n-gram language model is extracted from the TL monolingual corpus.

To that end, the free-to-use SRILM (Stolcke et al. 2011) package is used to create the model and to support the portability of this method to new language pairs. From the monolingual corpus, n-grams models are extracted in both token and lemma form. During the translation process, these can address a number of translation problems, including:

   i. the generation of tokens from lemmas during the post-processing phase;
   ii. the ordering of tokens within a phrase or near phrase boundaries;
   iii. the choice of the most appropriate translation of a word in TL, from a number of candidate translations.

Thus, the use of the n-gram model to test hypotheses helps to improve both the second phase of the translation process and the post-processing phase, as discussed in Tambouratzis et al. (2014). This new language model addresses the lack of training patterns from the parallel corpus, which are used to extract post-processing rules. Indicative examples that can be resolved include:

Article introduction and deletion: Given that there is no 1:1 mapping in the use of definite articles between languages, it is essential to check whether the definite article is correctly handled in TL (e.g. removed before proper names for English).

Generation of verb forms: Specific errors in the translation output include cases of active/passive voice mismatches between SL and TL. For example, the Greek deponent verb "ἔρχομαι" (come) (an active verb with mediopassive morphology) is translated as "be come" by the token generation component due to erroneously porting the verb's passive morphology from SL to TL.

In-phrase token order: The correct ordering of tokens within a given phrase occasionally fails to be established by the indexed corpus. Errors of this type can be remedied via the n-gram model.

Prepositional complements: When translating the prepositional complement of a verb (e.g. "depend + on"), it is likely that an incorrect preposition will be selected due to limited context information. Such a selection can be revised via the larger n-grams that incorporate contextual information.

6.2.1 Extracting the N-Gram Models

The additional effort needed to generate the n-gram LM for an existing PRESEMT translation system is limited. As the monolingual TL corpus is already lemmatised to establish the TL-indexed model, both lemma and token-based n-grams can be readily extracted, using the publicly available SRILM tool (Stolcke et al. 2011). Models have been extracted for values of n equal to 2 and 3, creating in total four n-gram models to support queries in factored representation levels. Witten-Bell smoothing is used and all n-grams with less than five occurrences over the monolingual corpus are filtered out to reduce the model size. Each n-gram model contains ca. 25 million entries, including the SRILM-derived logarithms of probabilities.

In PRESEMT, the indexed phrase model generates a first translation, upon which hypotheses are made that are tested for validity by retrieving the relevant n-gram information. If the n-gram model corroborates this hypothesis, no modification is applied, while if the n-gram-based likelihood leads to the hypothesis rejection (if an alternative translation is considered more likely based on the TLM), the translation is revised accordingly.

A set of hypotheses has been established based on the error analysis, for the language pair Greek to English. Each hypothesis is expressed by a mathematical formula which checks the likelihood of an n-gram, via either the lemma-based n-gram model or the token-based model. The decision is then made based on a threshold value (denoted as *thres_hi*, where *i* is the id.number of the hypothesis). The hypotheses used experimentally are summarised in Table 6.4. Each hypothesis is defined by specifying a set of assumptions (specific conditions that need to hold to trigger the hypothesis testing), then the formula to be checked and finally the replacement actions to be performed in the translation output.

Table 6.4 List of hypotheses used for the Greek to English language pair

Hypoth.	LM type	Aim	Definition of hypothesis	Thresh. value
Hyp1	2-gram lemma	Handling of deponent verbs (delete auxiliary "be")	ASSUMPTIONS: lem (A) = "be" & tag (B) = "VVN" Hyp1: $p(A, B) >$ thres_h1 IF Hyp1 rejected, THEN {A, B} → {B}	−4.50
Hyp2	2-gram lemma	Deletion of definitive article	ASSUMPTIONS: B = "the" with tag(B) = "DT" Hyp2: $(\min(p(A, B), p(B, C)) - p(A,C) <$ thres_h2 IF Hyp2 rejected, THEN {A, B, C} → {A, C}	+4.00
Hyp3	3-gram lemma	Deletion of preposition in case of consecutive ones	ASSUMPTIONS: tag (A) = "IN" & tag(B) = "IN" IF $p(A, C) > p(B, C)$ THEN {A, B, C} → {A, C} ELSE {A, B, C} → {B, C}	–
Hyp4	3-gram lemma	Selection of a more appropriate preposition	ASSUMP.: tag(D) = "IN" (D spans all prepositions) Hyp4: $p(A, B, C)-\max(p(A, D, C) >$ thres_h4 IF Hyp4 rejected, THEN {A, B, C} → {A, D, C}	+1.50
Hyp5	2-gram token	Replacement of tokens with respective lemmas	ASSUMPTIONS: None Hyp5: $p(A, B) >$ thres_h5 IF hyp5 rejected, THEN {A, B} → {lem(A), lem(B)}	+1.50
Hyp6	2-gram lemma	Deletion of the definitive article	ASSUMPTIONS: A == "the" with tag(A) = "DT" Hyp6: $p(A, B) >$ thres_h6 IF Hyp6 rejected, THEN {A, B} → {B}	−5.50

A, B and C represent consecutive words in the preliminary translation output on which the hypotheses are tested, and D is a new word to be introduced in the final translation

6.2.2 Experimental Results

The effectiveness of the proposed hypothesis checking on the different n-gram language models is presented here using as an example the Greek to English

Table 6.5 Comparative performance for evaluation dataset1, for different configurations with n-gram hypotheses, in comparison with the PRESEMT baseline (without n-gram LMs)

Configuration	BLEU (%)	NIST (%)	Meteor (%)	TER (%)
Baseline	0.00	0.00	0.00	0.00
Hyp1–4	0.00	+0.26	−0.18	−0.28
Hyp1–5	+0.65	+0.68	−0.01	−0.83
Hyp1–6	**+0.94**	**+1.26**	−0.33	**−1.56**

An improvement (deterioration) in BLEU, NIST and Meteor corresponds to positive (negative) values, while the inverse is true for TER

Table 6.6 Comparative performance for evaluation dataset2, for different configurations with n-gram hypotheses, in comparison with the PRESEMT baseline (without n-gram LMs)

Configuration	BLEU (%)	NIST (%)	Meteor (%)	TER (%)
Baseline	0.00	0.00	0.00	0.00
Hyp1–4	+1.70	+0.43	+0.15	−0.45
Hyp1–5	**+3.22**	+0.79	**+0.18**	−0.83
Hyp1–6	+2.92	**+1.56**	−0.01	**−1.03**

An improvement (deterioration) in BLEU, NIST and Meteor corresponds to positive (negative) values, while the inverse is true for TER

language pair. Out of the six hypotheses used, Hyp1 to Hyp4 are infrequently activated during translation, and thus when reporting the experimental results they are grouped together in one entry. On the contrary, the application of the more frequent hypotheses (Hyp5 and Hyp6) is reported separately. The scores obtained when hypotheses are activated are expressed as improvements over the baseline PRESEMT system that does not recourse to the n-gram model.

In the experiments, two datasets were used, namely dataset1 and dataset2 . In both cases, the introduction of hypotheses was compared to the baseline PRESEMT system, without any n-gram modelling (this being the baseline system). In Table 6.5, the improvements over the baseline PRESEMT system are reported for evaluation dataset1 when introducing the different hypotheses. Similarly, in Table 6.6, the improvements over the baseline PRESEMT system are reported for evaluation dataset2.

For dataset1 (cf. Table 6.5), the best BLEU score is obtained when activating all six hypotheses, and the same applies to NIST and TER. On the contrary, for Meteor the best result is obtained without resorting to the n-gram LM. Still the difference in Meteor scores is minor (less than 0.35%). The improvements in BLEU, NIST and TER are much more marked, being, respectively, +0.94, +1.2 and −1.5% over the baseline, when using all six hypotheses. Furthermore, as the number of activated hypotheses increases, each of these three scores improves monotonically.

The same experiment is repeated for dataset2 (cf. Table 6.6), to determine if the behaviour is consistent or if it is dataset-dependent. It turns out that the results are qualitatively similar. Again, the best scores for Meteor are obtained when no hypotheses are tested, though the score differences over differing levels of hypothesis activation are negligible. For the other metrics, though, recourse to the n-gram LM coupled with hypothesis testing results in an improvement to the scores obtained. The improvements obtained amount to just under 3.0% for BLEU, 1.6% for NIST and 1.0% for TER, over the baseline system scores indicating a measurable improvement. Notably, for dataset2, a higher BLEU score is obtained when all hypotheses but Hyp6 are activated.

A statistical analysis using a paired t-test has been undertaken to determine whether the additional n-gram modelling improves significantly the translation scores in the case of dataset2. It turns out that though no statistically significant difference was detected for any metric at a 0.05 confidence level, significant improvements were detected for BLEU, NIST and TER at a 0.10 confidence level. Thus, a meaningful improvement in translation accuracy is achieved.

6.2.3 Discussion

Experimental results indicate that n-gram models can correct errors in the translation output, contributing to an improved translation quality. This is verified for a well-developed state of the Greek to English language pair, indicating that the introduction of hypothesis testing via n-gram LMs results in meaningful improvements in the objective evaluation metrics. It is likely that further improvements in objective metrics can be achieved by expanding the suite of hypotheses. A possible improvement would involve, for example, supplementing the lexical information with detailed PoS information.

Usefully, the new n-gram modelling requires no collection of additional monolingual corpora or further annotation beyond what has already been performed for the basic PRESEMT system. The n-gram database is created via an established software package (SRILM), via which hypotheses are rejected or confirmed. This results in a straightforward implementation.

Tests performed so far have used empirically set threshold values for the hypotheses. Therefore, it is likely that more appropriate threshold values may be established. A systematic methodology such as MERT (Och 2003) or genetic algorithms (Sofianopoulos and Tambouratzis 2010) could be employed to optimise the actual values of the thresholds.

References

Black PE (2005) Dictionary of algorithms and data structures. U.S. National Institute of Standards and Technology (NIST)

Brown PF, Della Pietra SA, Della Pietra VJ, Mercer RL (1993) The Mathematics of Statistical Machine Translation: Parameter Estimation. Computational Linguistics 19(2):263–311

Duda RO, Hart PE, Stork DG (2001) Pattern classification, 2nd edn. Wiley Interscience, New York

Hwa R, Resnik P, Weinberg A, Cabezas C, Kolak O (2005) Bootstrapping parsers via syntactic projections across parallel texts. Nat Lang Eng 11:311–325

Koehn P, Hoang H (2007) Factored translation models. In: Proceedings of the 2007 Joint Conference on Empirical Methods in Natural Language Processing and Computational Natural Language Learning, Prague, Czech Republic, pp 868–876

Lafferty J, McCallum A, Pereira F (2001) Conditional random fields: probabilistic models for segmenting and labelling sequence data. In: Proceedings of ICML Conference, 28 June–1 July, Williamstown, USA, pp 282–289

Och FJ (2003) Minimum error rate training for statistical machine translation. In: Proceedings of the 41st Annual Meeting of the Association for Computational Linguistics (ACL), Sapporo, Japan, July, pp 160–167

Sha F, Pereira FCN (2003) Shallow parsing with conditional random fields. In: Proceedings of HLT-NAACL Conference, pp 213–220

Smith DA, Eisner J (2009) Parser Adaptation and Projection with Quasi-Synchronous Grammar Features. In: Proceedings of the 2009 Conference on Empirical Methods in Natural Language Processing, Singapore, vol 2, pp 822–831

Sofianopoulos S, Tambouratzis G (2010) Multiobjective optimisation of real-valued parameters of a hybrid MT system using Genetic Algorithms. Pattern Recogn Lett 31(12):1672–1682

Stolcke A, Zheng J, Wang W, Abrash V (2011) SRILM at sixteen: update and outlook. In: Proceedings of IEEE Automatic Speech Recognition and Understanding Workshop, December 2011

Tambouratzis G (2014) Comparing CRF and template-matching in phrasing tasks within a Hybrid MT system. In: Proceedings of the 3rd Workshop on Hybrid Approaches to Translation (held within the EACL-2014 Conference), April 27, Gothenburg, Sweden, pp 7–14

Tambouratzis G, Simistira F, Sofianopoulos S, Tsimboukakis N, Vassiliou M (2011) A resource-light phrase scheme for language-portable MT. In: Proceedings of the 15th International Conference of the European Association for Machine Translation, 30–31 May, Leuven, Belgium, pp 185–192

Tambouratzis G, Sofianopoulos S, Vassiliou M (2014) Expanding the Language model in a low-resource hybrid MT system. In: Proceedings of SSST-8 Workshop, held within EMNLP-2014, 25 October 2014, Doha, Qatar, pp 57–66. ISBN 978-1-937284-96-1

Tsuruoka Y, Tsujii J, Ananiadou S (2009) Fast full parsing by linear-chain conditional random fields. In: Proceedings of the 12th Conference of the European Chapter of the ACL, Athens, Greece, 30 March–3 April, pp 790–798

Wallach HM (2004) Conditional random fields: an introduction. CIS Technical Report, MS-CIS-04-21. 24 February 2004, University of Pennsylvania

Yarowsky D, Ngai G (2001) Inducing multilingual POS taggers and NP bracketers via robust projection across aligned corpora. In: Proceedings of NAACL-2001 Conference, pp 200–207

Chapter 7
Conclusions and Future Work

This chapter performs a review of the research work discussed in the previous chapters of the present volume. This review represents a summary of the outcomes of the research within the PRESEMT project. As a logical outcome, a set of key directions is identified for future work in order to further improve the MT methodology. A brief report of the most promising ones is provided in the second part of this chapter.

7.1 Review of the Effectiveness of the PRESEMT Methodology

As discussed in earlier chapters, the proposed methodology can develop translation systems for arbitrary language pairs. It has been shown to be relatively easy to implement, requiring only limited amounts of specialised resources and a limited amount of effort to set up a working MT system. In particular, the amount of parallel corpora required is very low, in accordance with the specifications set at the beginning of the development. This provides the proposed method with a substantial advantage in comparison with Statistical MT systems. Typically, the experiments carried out have used less than 500 sentences of parallel corpus in SL and TL, which are aligned at sentence level. Such sizes of parallel corpus can definitely be compiled by a single person in a couple of days, even if it is impossible to retrieve a parallel corpus from other sources.

To counterbalance the very small size of the parallel corpus, the use of monolingual corpora is widespread in PRESEMT, improving the applicability of the proposed method. Furthermore, as monolingual corpora are abundant over the Web, it is possible to choose appropriate subsets to cover a specific domain. Thus, in contrast to parallel corpora, monolingual ones can be used in a virtually unconstrained manner, in an effort to maximise the translation accuracy.

© The Author(s) 2017
G. Tambouratzis et al., *Machine Translation with Minimal Reliance on Parallel Resources*, SpringerBriefs in Statistics,
DOI 10.1007/978-3-319-63107-3_7

Other resources are required for PRESEMT in order to create a new MT system, such as a shallow parser for the TL as well as a bilingual dictionary from SL to TL. Efforts have been made to reduce the knowledge explicitly provided and instead develop and integrate tools for extracting knowledge in an automated manner. Thus, within the free-to-download and use PRESEMT package, suites of tools are provided. These tools cover extracting language models from the monolingual corpus, establishing structural transformations from the parallel corpus and establishing the phrasing scheme in SL by porting the phrasing of a parser from TL.

In terms of performance, experiments have shown that the PRESEMT performance is not as good as leading MT systems, such as Bing Translator and Google Translate. This can be expected as the amount of specialised information explicitly provided in the system is much less. For instance, it is acknowledged that MT systems such as those of Bing and Google liberally use a number of linguistic resources in order to achieve the optimum translation performance. This is of course fully justified within the quest for a translation quality which is as high as possible.

On the other hand, the adoption of such an approach necessitates the availability of a wealth of resources, coupled with a large development team. In addition, it is highly likely that different algorithmic solutions will be used for different language pairs, while making use of dedicated linguistic tools that are finely tuned for each specific language. The minimalist approach of PRESEMT, on the other hand, in our view represents a realistic set-up to cover more language pairs while requiring the investment of minimal resources.

The PRESEMT project has been aimed to evaluate the feasibility of such a methodology. As a rule, it has been seen that continued development effort can lead to substantial improvements in the translation performance. This development extends over both the lexicon coverage and the modules that extract the algorithmic resources. Having reached a stable point in the development, the emphasis here is placed on performing a review and determining whether the performance of the PRESEMT methodology can be improved further as compared to its current state. This is tackled in the next subsection, where a number of possible developments are covered.

7.2 Likely Avenues for Improvements in Translation Quality

During the development of the PRESEMT methodology covered in this book, a number of weaknesses have been identified based on an analysis of the translation output and the errors contained. Based on these, and taking into account relevant research in the area of Machine Translation and natural language processing, a number of future developments have been identified, which cover improvements in both the linguistic resources and the algorithms used. These are itemised below.

7.2.1 Automatic Enrichment of Dictionary

The bilingual dictionary between source and target languages is a relatively expensive resource, which in several cases will have to be sourced from an external provider. This means that the dictionary may be either out-of-date or incomplete, thus being sub-optimally matched to the domain on which the MT system will be applied. The question then is how to best augment this dictionary, without investing a substantial number of person-months, implying that an algorithmic solution is required. In the literature, a number of methods for enriching dictionaries from parallel or near-parallel corpora have been proposed. Earlier work which is of relevance includes that of, e.g. Koehn and Knight (2002) in extending an existing lexicon. More recent work has included extracting lexical translation probabilities (Klementiev et al. 2012) from monolingual corpora. In addition, the use of non-parallel monolingual corpora in both SL and TL is studied by Nuhn et al. (2012).

7.2.2 Design and Implementation of TEM

Currently, disambiguation in PRESEMT is performed within phase 2 of the translation process, namely the Translation Equivalent Selection. This is, however, sub-optimal as both the sequencing of tokens and the disambiguation are performed concurrently, by accessing the indexed TL model extracted from the parallel corpus. Error analysis on the translation output has shown that it would be preferable to use a dedicated word sense disambiguation module to support this process and thus improve the translation accuracy.

Within the PRESEMT project lifetime, substantial effort was invested on word sense disambiguation. Three different approaches were investigated:

- **Vector Space models** (VSM) (Lynum et al. 2012) proved to be computationally very heavy and did not give the expected improvement in performance;
- **Classic n-gram models** were trained on the TL corpus, but the resulting n-gram model had limited effectiveness;
- **Self-Organising Maps** (SOM) (Kohonen 1997) were applied to create topology-preserving mappings of possible TL tokens, without, however, achieving the desired improvements in accuracy.

The common aspect of the aforementioned methods was that their results were of limited effectiveness. In addition, though these models may be created in an offline process, the resulting models should be accessed in an efficient manner during run-time and without posing excessive requirements to processing power. In this respect, solutions such as VSMs are less than ideal. One possibility would be to create paired models of SL and TL in order to model the information from both SL and TL in combination, to improve the translation performance.

Another option would be to revisit SOM-based methods. SOM is a computational intelligence-based method for creating topology-preserving mappings of complex data (i.e. maps where the distance in pattern space is projected to a low-dimensional lattice), which has been successfully applied to a diverse set of pattern spaces, including textual data, numerical data, images. SOM has been used for self-organisation of massive collections of texts operationally (Kohonen et al. 2000). One likely source for improvements regarding MT is to use dynamically growing SOM-based models to create more representative semantic models, the added benefit being that SOM-based models operate at a much reduced dimensionality thus simplifying calculations.

Also, less rigid architectures of self-organising systems with adaptive structures have been proposed, including the Growing Neural Gas networks (Fritzke 1995) and Growing Hierarchical SOM (Rauber et al. 2002), which have the ability to closely fit the map to the pattern space characteristics, by amending the size and dimensions of the map. It is believed that these latter methodologies can generate topological maps for the words of a language that indicate semantic similarity via a reduced distance in the network lattice.

These low-dimensional maps could then be used during run-time to rapidly determine the best translations of a given SL word in TL. Here, the flexibility of the network structure in comparison with other works is of particular importance in achieving a more effective disambiguation, as the structure may be adapted to the specific language (and even domain) used.

7.2.3 Grouping of Tokens and PoS Tags into Related Classes

The size of the data collected, in particular the bilingual corpus, is relatively limited. The question, thus, becomes how the system can better augment this data and handle previously unseen entries. For instance, for the Greek-to-English language pair, most of the tag characteristics are not taken into account. When matching patterns in SL, aspects such as the number or gender are omitted and only the Part-of-Speech and case are retained during phase 1 of the translation process (Structure Selection).

However, it could be advantageous to further cluster together the main PoS categories to augment the training data. For instance, nouns and pronouns can be used interchangeably within the formation of noun chunks (NC). Hence, the a priori provision of such information would allow PRESEMT to create more accurate models of the different chunk types, or model phrase types more comprehensively. This direction follows a general tendency as exemplified for instance by the proposal for a universal tagset (Petrov et al. 2012) and is also along the lines of the work of Stymne (2009). The main contribution is evidently that it becomes possible to reduce the pattern space and thus probably train more effectively different system modules, ranging from the phrasing model (PMG) to the first step of the translation process.

7.2.4 Revision of the Structure Selection Translation Phase

Studies of the PRESEMT-generated translations have shown that the chosen TL structures are in several cases sub-optimal. This is due to the fact that the parallel corpus, from which the translation structure can be deduced, is very limited.

Based on this, one possibility would be to try to bring closer together the SL and TL languages, in terms of structure, drawing inspiration from similar work in SMT systems. This would imply a type of pre-ordering, to determine frequent reorderings and to improve performance when translating between languages where many reorderings are needed, as proposed by Xia and McCord (2004). Recently, increased research effort has been invested in pre-ordering algorithms using different techniques, including discriminative classifiers (Lerner and Petrov 2013), neural networks (Gispert et al. 2015) and word class clustering (Stymne 2012).

Similar efforts could be applied to the PRESEMT methodology. The key requirement to performing pre-ordering is to have a dependency parser available in SL. This is a sophisticated tool that may not be expected to be easily available in lesser-resourced languages such as those considered for the PRESEMT methodology. Therefore, in such cases, alternative methods to define the Structure Selection are required, which should exploit more effectively the small parallel corpus available.

One approach which conceptually resembles the pre-ordering mechanisms proposed for SMT is to search through the parallel corpus to establish templates of, e.g. tag sequences for which realignments (amendments in the sequence of phrases between SL and TL) occur frequently. Hence, a list of possible realignments will be established for application during translation. As an example, if the SL side phrase sequence {A; B; C} is frequently translated into the structure {B'; C'; A'} in TL, this structure modification is stored. The result is that instead of determining the structure of an entire sentence in one step by comparing it to the few known sentence structures (at most 200 for a 200-sentence parallel corpus) with an EBMT-type approach, the structure-definition process is split into smaller steps. In each such step, a match to one sub-sentential template is sought, and if this is achieved, then localised realignments are performed. Thus, the SL structure is modified only when localised realignments are found to be applicable.

As an alternative, a template-matching approach (Black 2005) is based on the underlying assumption that changes in structure are generally localised. In this approach, longer-range reorderings may also be established, (for instance in English-to-German, where the verb participle is always positioned at the end of the main clause). Though initial experiments have shown that realignment templates can successfully identify such reorderings, more investigation is required to determine the accuracy of this method, especially in the absence of very extensive training data.

Experiments with this approach have recently been found to result in improved translation accuracy in comparison with the standard Structure Selection algorithm of PRESEMT (Tambouratzis and Pouli 2015). More specifically, when tested over two datasets, an improvement of between 1 and 2% has been achieved for BLEU, while for TER an improvement of around 1% is recorded. For NIST and Meteor, the improvements achieved are more modest.

7.2.5 Improving the Alignment of Words/Phrases

The parallel corpus is the main source of information on structural transformations from SL to TL, based on the SL-TL alignment of words and phrases yielded by PAM. The extrapolation of phrases for arbitrary input text is achieved by the PMG module, which is trained on the alignments of words as established by PAM. As both PAM and PMG are applied at the very beginning of the translation process, any errors in either module will propagate through the translation chain and multiply. Hence, by improving either PAM or PMG, the translation accuracy can be expected to benefit.

PAM is rule-based with a strict linear set of steps. It is believed that fuzzy-matching techniques applied iteratively could be employed to improve the PAM performance. On the other hand, in the default configuration PMG is based on Conditional Random Fields, which implement higher order models which need to be trained sufficiently. It is expected that the limited training set may inhibit the accuracy of the finally evolved PMG. It is likely that alternative methods based on computational intelligence may provide more accurate segmentation results.

7.2.6 Augmenting the TL Language Model to Cover Supra-Phrasal Segments

A final area for improvements is that of the basic PRESEMT TL language model, which is created from the very large monolingual corpus. The question is whether it is possible to increase the coverage of the indexed files as compared to the basic configuration discussed in Chap. 3. Sufficient training data exists in the form of monolingual corpora, and what is required is to process them efficiently in order to extract information which is found to be missing during the translation process.

In this case, improvements are bound to be developed on the basis of experimental results. An error analysis procedure has shown that larger phrases (in terms of the number of tokens contained) are more difficult to translate accurately, most frequently resulting in an erroneous ordering of tokens in the translation output. In this case, a favoured solution is to introduce in the indexed corpus patterns representing larger phrases, by, for example, providing merged sequences of consecutive phrases. The idea underlying this approach is that the system becomes more tolerant to non-standard segmentations of the text to be translated into larger phrases. Preliminary results at the time of finalising this volume have shown a small but marked improvement in terms of the translation accuracy (around 1% in terms of BLEU score). A further benefit has been found to be the ability to achieve a more consistent translation accuracy, irrespective of the actual parsing scheme used for the SL side language, which is attributable to the more complete coverage of the phrasing model.

Similar approaches have been already proposed in the literature, under the term of growing models. In many natural language processing tasks including speech recognition and more recently Machine Translation, statistical language models are predominantly extracted from n-grams, comprising exactly n consecutive tokens. It has been acknowledged that n-gram models where n is fixed do not provide a perfect model of the language, but that improvements are possible by employing models with larger sizes of n, this giving rise to variable n-gram language modelling (Siu and Ostendorf 2000) and subsequently growing language models (Siivola and Pellom 2005). Recently, language model growing has been shown to improve performance in MT applications as well (Wang et al. 2014).

It is claimed that similar improvements can be achieved if the indexed model used by PRESEMT for the target language is augmented, e.g. by injecting in the model sequences of consecutive phrases as single phrases. This injection will provide appropriate templates of how these larger sequences of tokens should be ordered. Preliminary experiments have shown that this proposed enrichment of the language model can give substantial improvements, for all four objective metrics studied. The largest improvements are achieved for BLEU and typically range between 4 and 6% (Tambouratzis and Pouli 2016).

7.2.7 A Closing Evaluation of Translation Accuracy

Instead of an epilogue, it has been decided to provide a final comparison between two MT methodologies, both of which are provided as free-to-use Corpus-Based MT methodologies. This indicative example contains the most recent results of PRESEMT (Tambouratzis and Pouli 2015), obtained in late June 2015 as this concluding chapter was being finalised. In the case of PRESEMT, these latest results were obtained on the standard PRESEMT 200-sentence development set, using a single-reference evaluation set for the Greek-to-English language pair. For this, the PRESEMT-derived system gave translations with BLEU and NIST scores of 0.3626 and 7.086. As a baseline, a Greek-to-English Moses-based system was used, which was trained with a parallel corpus of approx. 1.2 million sentences. Thus, Moses has a parallel corpus four orders of magnitude larger than that used by PRESEMT, and resulted in BLEU and NIST scores of 0.3795 and 7.039, respectively.

The behaviour of the two systems is closely comparable, indicating the promising nature of the PRESEMT methodology in general, even when compared to a widely used methodology such as Moses. It should be noted that both

PRESEMT and Moses use algorithmic solutions to extract knowledge from corpora from which to train a working MT and both provide the code for use by other research groups. Both have also been supported by the European Commission research funding agencies. Of course, Moses has been developed more extensively, over a longer time span and via a series of projects. In addition, it has been researched over substantially more language pairs. On the other hand, the requirements it poses over the resources required to develop an MT system are higher than that of PRESEMT. PRESEMT requires a small parallel corpus, while Moses requires a considerably larger one. One could argue that the Moses parallel corpus need be less strictly parallel, in comparison with the PRESEMT one, as its size allows it to be more fault-tolerant. On the other hand, it is arguable that the PRESEMT parallel corpus can be hand-checked in a very limited time frame, while for Moses this is not a realistic option. Also, PRESEMT requires an existing bilingual dictionary between SL and TL, while Moses extracts this automatically from the parallel corpora.

However, for language pairs for which very limited or no appropriate resources (parallel corpora) are readily available, PRESEMT appears to be a better match than Moses (or more generally SMT systems) for developing a new system, without sacrificing a lot in terms of translation accuracy.

References

Black PE (2005) Dictionary of algorithms and data structures. U.S. National Institute of Standards and Technology (NIST)

Fritzke B (1995) A growing neural gas network learns topologies. In: Advances in neural information processing systems, vol 7. MIT Press, Cambridge, pp 625–632

Gispert A, Iglesias G, Byrne B (2015) Fast and accurate preordering for SMT using neural networks. In: NAACL-HLT-2015 Conference

Klementiev A, Irvine A, Callison-Burch C, Yarowsky D (2012) Toward statistical machine translation without parallel corpora. In: Proceedings of EACL-2012, Avignon, France, 23–25 April, pp 130–140

Koehn P, Knight K (2002) Learning a translation lexicon from monolingual corpora. In: ACL Workshop on Unsupervised Lexical Acquisition

Kohonen T (1997) Self-Organising Maps (2nd ed.) Berlin, Springer-Verlag

Kohonen T, Kaski S, Lagus K, Salojarvi J, Honkela J, Paatero V, Saarela A (2000) Self organisation of a massive document collection. IEEE Trans Neural Networks 11(3):574–585

Lerner U, Petrov S (2013) Source-side classifier preordering for machine translation. In: Proceedings of EMNLP-2013 Conference, Seattle, USA, October 2013, pp 513–523

Lynum A, Marsi E, Bungum L, Gambäck B (2012) Disambiguating word translations with target language models. In: Text, speech and dialogue: Proceedings of the TSD-2012 International Conference, Brno, Czech Republic, 3–7 September. Springer, Brno, pp 378–385

Nuhn M, Mauser A, Ney H (2012) Deciphering foreign language by combining language models and context vectors. In: Proceedings of the ACL-2012 Conference, Jeju, Republic of Korea, 8–14 July, pp 156–1643

Petrov S, Das D, McDonald R (2012) A universal part-of-speech tagset. In: Proceedings of the 8th International Conference on Language Resources and Evaluation (LREC '12), 23–25 May, Istanbul, Turkey

Rauber A, Merkl D, Dittenbach M (2002) The growing hierarchical self-organizing map: exploratory analysis of high-dimensional data. IEEE Trans Neural Networks 13(6):1331–1341

Siivola V, Pellom BL (2005) Growing an n-gram language model. In Proceedings of Interspeech'05 Conference, Lisbon, Portugal, pp 1309–1312

Siu M, Ostendorf M (2000) Variable n-gram language modeling and extensions for conversational speech. IEEE Trans Speech Audio Process 8(1):63–75

Stymne S (2009) A comparison of merging strategies for translation of German compounds. In: Proceedings of EACL, Student Research Workshop, Athens, Greece, pp 61–69

Stymne S (2012) Clustered word classes for preordering in statistical machine translation. In: Proceedings of the 13th EACL Conference, 23–27 April, Avignon, France, pp 28–34

Tambouratzis G, Pouli V (2015) Establishing sentential structure via realignments from small parallel corpora. In: Proceedings of HYTRA-2015 Workshop, held within ACL/IJCNLP-2015, Beijing, China, 31 July, pp 21–29

Tambouratzis G, Pouli V (2016) Linguistically inspired language model augmentation for MT. In: Proceedings of LREC-2016, 23-28 May 2016, Portoroz, Slovenia. ISBN 978-2-9517408-9-1

Wang R, Zhao H, Lu B-L, Utiyama M, Sumita E (2014) Neural network based bilingual language model growing for statistical machine translation. In: Proceedings of the EMNLP-2014 Conference, Doha, Qatar, 25–29 October, pp 189–195

Xia F, McCord M (2004) Improving a statistical MT system with automatically learned rewrite patterns. In: Proceedings of Coling 2004, Geneva, Switzerland, August 23–27, pp 508–514

Glossary

The following table summarises the acronyms used throughout this volume, ordered alphabetically, for ease of reference.

ADJC ADJectival Chunk

ADVC ADVerbial Chunk

ALPAC Automatic Language Processing Advisory Committee

BLEU Bilingual Evaluation Understudy

CRF Conditional Random Fields

CONJC CONJunction Chunk

CS Corpus Sentence

CBMT Corpus-Based Machine Translation

EBMT Example-Based Machine Translation

XML Extensible Markup Language

fhead Functional Head

HMM Hidden Markov Models

HMT Hybrid MT

Hyp Hypothesis

IS Input Sentence

ISP Input Sentence Phrase

INTJ Interjection Chunk

ISC Isolated word Chunk

LM Language Model

MT Machine Translation

© The Author(s) 2017
G. Tambouratzis et al., *Machine Translation with Minimal Reliance on Parallel Resources*, SpringerBriefs in Statistics,
DOI 10.1007/978-3-319-63107-3

ME Maximum Entropy

Meteor Metric for Evaluation of Translation with Explicit ORdering

MCP Monolingual Corpus Phrase

NNS Non-Null subject

NC Noun Chunk

NS Null-subject

OOV Out-Of-Vocabulary

PRT Particle Chunk

PoS Part-of-Speech

PRESEMT Pattern Recognition-Based Statistically Enhanced MT

PAM Phrase Aligner Module

PMG Phrasing Model Generation

PC Prepositional Chunk

RBMT Rule-Based Machine Translation

SOM Self-Organising Map

SL Source Language

SMT Statistical Machine Translation

SS Structure Selection

TL Target Language

TLM Target Language Model

TEM Template Matching

TES Translation Equivalent Selection

TER Translation Error Rate

VSM Vector Space Model

VC Verb Chunk

Printed in the United States
By Bookmasters